# of Wyoming

David Lageson
Darwin Spearing

MOUNTAIN PRESS PUBLISHING COMPANY
MISSOULA 1988

Revised 2nd Edition, March 1991

Seventh Printing, November 2000

Library of Congress Cataloging-in-Publication Data

Lageson, David R.
Roadside geology of Wyoming.

(Roadside geology series)
Bibliography: p.
Includes index.
1. Geology—Wyoming—Guide-books 2. Wyoming
Description and travel—1981- —Guide-books.
I. Spearing, Darwin. II. Title III. Series.
QE181.L34 1988 557.87 86-1650
ISBN 0-87842-216-1 (pbk.)

Mountain Press Publishing Company
P.O. Box 2399 • Missoula, MT 59806
1-800-234-5308

Acknowledgements

We wish to acknowledge collectively the contributions of all the geologists who have helped to unravel the complex and exciting geological history of Wyoming. The Wyoming Geological Association, Geological Survey of Wyoming, and United States Geological Survey provided many publications that were helpful in the compilation of this book. Selected references are provided at the end of the book along with information sources, glossary, and a list of museums in Wyoming with geological information. Regrettably, in a book of this type, we cannot cite the published work of all geologists who have worked in Wyoming. David Alt and Donald Hyndman at the University of Montana carefully reviewed the book, and we express our thanks to them for their helpful suggestions. We also express our appreciation to Diane Lageson for her assistance with historical research, writing, and proofreading.

All photographs were taken by the authors, except for those on p. 129, 131 - originals by W.H. Jackson, U.S. Geological Survey 1870, from prints courtesy of Fort Caspar Museum, Casper, Wyoming; p. 72, 84 - courtesy of National Archives, Washington, D.C.; p. 247 - courtesy of Montana Travel Promotion Unit, Helena, Montana.

All illustrations, including the cover design, were hand drawn by the authors either from the authors' conceptions or modified from the sources indicated in captions. Unless otherwise noted, road maps were hand drawn from the Geologic Map of Wyoming, 1985, U.S. Geological Survey, J.D. Love and A.C. Christiansen. A complete list of references is located at the back of the book.

Contents

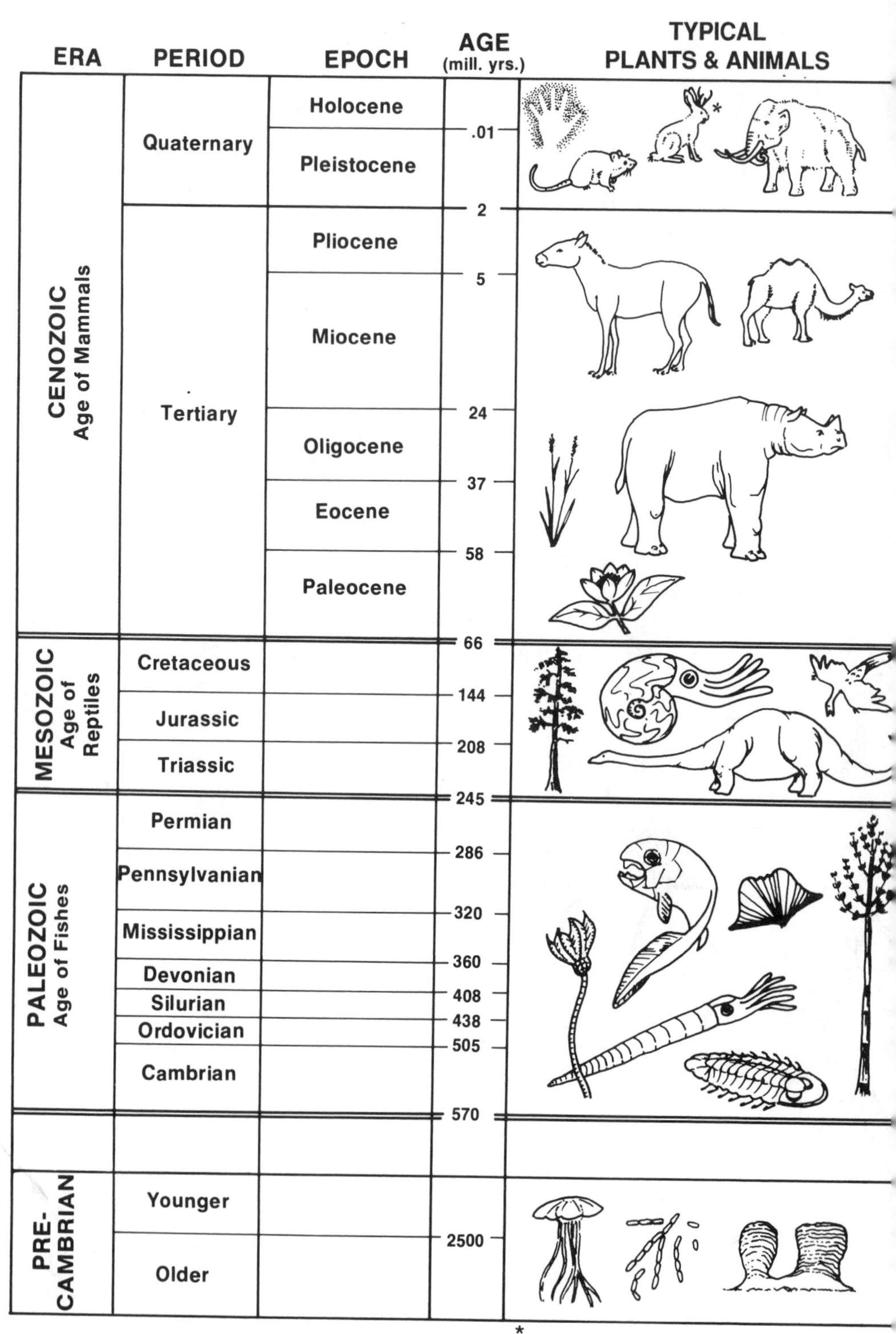

ERA	PERIOD	EPOCH	AGE (mill. yrs.)	TYPICAL PLANTS & ANIMALS
CENOZOIC Age of Mammals	Quaternary	Holocene		
			.01	
		Pleistocene		*
			2	
	Tertiary	Pliocene		
			5	
		Miocene		
			24	
		Oligocene		
			37	
		Eocene		
			58	
		Paleocene		
			66	
MESOZOIC Age of Reptiles	Cretaceous			
			144	
	Jurassic			
			208	
	Triassic			
			245	
PALEOZOIC Age of Fishes	Permian			
			286	
	Pennsylvanian			
			320	
	Mississippian			
			360	
	Devonian			
			408	
	Silurian			
			438	
	Ordovician			
			505	
	Cambrian			
			570	
PRE-CAMBRIAN	Younger			
			2500	
	Older			

* The jackalope, a mythical Wyoming creature.

— MAJOR EVENTS —

Explosive eruption of Yellowstone (3 times); uplift of the Teton Range; regional uplift of the western U.S. with concurrent canyon-cutting and excavation of older Laramide basins and mountains; at least 3 major glaciations in last 200,000 years (Buffalo, Bull Lake, Pinedale).

Continued basin filling; rivers flow unhindered across buried mountain ranges. Landscape is semi-flat and unremarkable.

Continued basin filling; mammals dominate the land; Miocene and Pliocene strata are preserved in central Wyoming's Sweetwater Hills (also called Granite Mountain).

Extensive filling of structural basins by sediments eroded from mountains. Rivers are sluggish with volcanic ash, blown in from west. Deposition of ash-rich White River formation.

End of Laramide mountain-building episode. Major coal, oil shale, and trona deposits. Lakes and fish in Green River Basin. Heart Mountain detachment forms in Beartooth Mountains. Eruption of Absaroka volcanic field.

Laramide orogeny - development of mountains and basins across Wyoming. Fort Union coal beds deposited in Powder River Basin.

Major mountain building begins to the west; deposition of gray shales and sands across Wyoming in "western interior seaway." Extinction of the dinosaurs.

Sundance seaway covers state with deposition of oyster-bearing sandstone, followed by red and green, river deposited sediments of the Morrison formation. Dinosaurs found in Morrison.

Extensive deposition of red sands (Chugwater formation); early dinosaurs.

Upwelling sea currents deposit phosphate in west; red sands and shales deposited to east; global extinction of many marine invertebrate animals.

Wind-blown sand blankets the state (Tensleep, Weber, Casper, Minnelusa formations); oil and gas later accumulates in these sands.

Shallow, warm sea covers the entire state; deposition of highly fossiliferous Madison limestone.

Uplift in southeast Wyoming; diamond-bearing diatremes intruded; limestone deposits in shallow seas to west (Darby & Jefferson formations).

Silurian limestones were deposited and then eroded prior to Devonian; Erosion break.

Shallow sea with deposition of Bighorn dolomite.

Widespread marine seas covered Wyoming repeatedly during the Paleozoic, starting with the Cambrian. Flathead sandstone was a beach deposited across the old, eroded Precambrian surface. Beginning of well-preserved, hard-shelled fossils of all phyla.

Mullen Creek-Nash Fork shear zone forms in southeast Wyoming due to a continent-continent collision 1.7 billion years ago. Sherman granite intruded near Laramie 1.4 billion years ago. Black dike on Mount Moran (Teton Range) intruded about 1 billion years ago. First evidence of soft-bodied animals and algae.

Sediments are deposited over 3 billion years ago, then are deeply buried and metamorphosed by heat and pressure into gneiss and schist during an ancient mountain-building episode. Granitic bodies intrude these older metamorphic rocks. Gneiss, schist and granite are now exposed in the core of younger mountains like the Wind River Range.

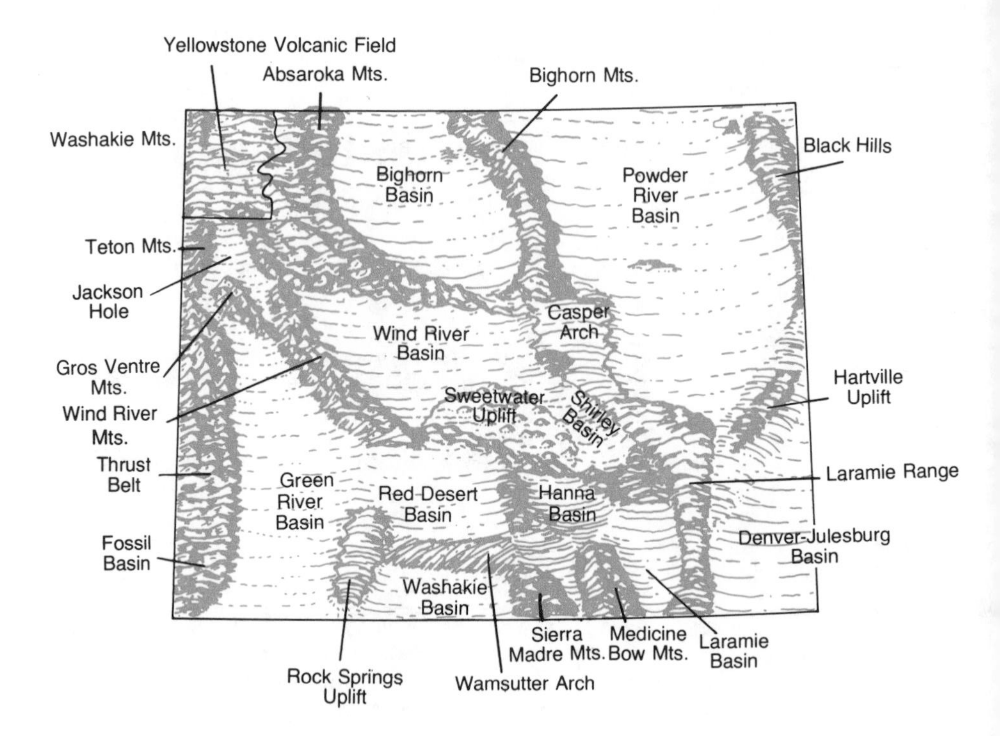

Mountains and basins of Wyoming.

Roadside Geology of Wyoming

Introduction

Above all else, Wyoming is a geological state. Wyoming may be known as the "Cowboy State," and the license plates display a cowboy on a bucking bronco, but to our minds, Wyoming is first and foremost geological! In every corner of every mountain range and basin within this big state one can find a geological story recorded in the rocks. In some places the story reads clear, in others whole chapters are missing, but that is the nature of geology. Geologists piece together incomplete puzzles about the history of the Earth. Our purpose in writing this book is to convey the overall story of Wyoming's geology and history based on the clues left in her rocks.

For both authors, Wyoming represents the essence of Rocky Mountain geology. As young geologists we "cut our teeth" on Wyoming's geology and our professional careers have been entwined with Wyoming ever since. For us, Wyoming is a "feeling," unique to this part of the Rocky Mountains. It is the feeling of broad, sage-covered basins extending to the horizon, and being able to see for 100 miles in any direction without a house, farm, road, or tree to obstruct the view. It is the feeling of wind, always present, gathering speed through the open basins and funneling around the mountains. It is the feeling of early evening in Jackson Hole when the Tetons cast long shadows across the valley floor. It is the feeling of antelope grazing on spring grass along the Medicine Bow River.

John Wesley Powell, famous geologist and explorer of the Colorado River drainage, and early director of the U.S. Geological Survey, wrote about the landscape of southwestern Wyoming in 1869 when he started down the Green River on his epic float trip through the Grand Canyon. Powell captured in words the "feeling" of Wyoming better than we are able to:

> Standing on a high point, I can look off in every direction over a vast landscape, with silent rocks and cliffs glittering in the evening sun. Dark shadows are gathering in the valleys and gulches, and the heights are made higher and the depths deeper by the glamour and witchery of light and shade. Away to the south the Uinta Mountains stretch in a long line, —high peaks thrust into the sky, and snow fields glittering like lakes of molten silver, and pine forests in somber green, and rosy clouds playing around the borders of huge, black masses; and heights and clouds and mountains and snow fields and forests and rocklands are blended into one grand view. Now the sun goes down, and I return to camp.

This book was written for those who would like to know more about the geology of Wyoming, including Yellowstone and Grand Teton national parks, but who know little or nothing about geology. We did not write this book for our professional colleagues in geology, although we hope that they, too, will enjoy it. Don't be afraid to plunge into the book; it won't bite! All the technical words have been defined in the text or glossary at the end of the book, and the many illustrations will help you visualize the geology. We believe that your appreciation and enjoyment of this great state will be enhanced by "rubbing elbows" with the rocks and history as you drive through.

HUMAN HISTORY AND GEOLOGY

Human habitation and cultural development are profoundly affected by the landscape and natural resources available, and these are the direct product of geology. Wyoming's history of emigration, settlement, resource and agricultural development have been directly controlled by geological factors.

The Indians, of course, were the first natives of Wyoming, although they too migrated to western North America from Asia several thousands of years ago, probably across the Bering Strait. The Indians of Wyoming were nomadic, following the great herds of buffalo, hunting bighorn sheep and elk in the high mountains, and wintering along the banks of the big rivers. Various tribes occupied the region, including the Crow, Shoshone, Sioux and Cheyenne.

The white man came into Wyoming in the early 1800s. The Louisiana Purchase of 1804 gave federal sanction to explorers and fur trappers to open the west for development and settlement. The first expedition to cross Wyoming was the Astorians in 1811-1812. Fur trappers and traders, seeking sleek beaver pelts to be made into fashionable hats for gentlemen, explored every nook and cranny of the state, and Wyoming is rich with their history and namesakes: Jackson's Hole, Colter, Leigh and Jenny lakes, Black's Fork, Smith's Fork, Fontenelle, Sublette, Bonneville, Rendezvous Peak, Bridger, and many others. The mountain men gathered each summer at a predetermined place for a rendezvous to trade furs for supplies, catch-up on old news and, in general, enjoy the comradery of men in a land with no laws! Famous rendezvous sites in Wyoming include the Green River Rendezvous (1833, 1835, 1837, 1839, and 1840) north of Pinedale, the Ham's Fork Rendezvous (1834), and the Wind River Rendezvous (1830 and 1838) near Riverton.

The geological landscape of Wyoming served as a thoroughfare for western migration over the last 150 years—people always passing through but rarely staying. For the tens of thousands of pioneers on the Oregon Trail in the 1840s to 1860s, Wyoming was the means to a better life in the Pacific Northwest, California, or Utah, but it was no place to stop. To the south lay the impenetrable walls of the Colorado Rockies, to the north lay equally forbidding mountains in Montana and Idaho. The "Wyoming Basin" provided a natural pass for the Conestoga wagons across the imposing backbone of the Rocky Mountains.

When the Transcontinental Railroad was built in the 1860s, Wyoming was again the natural choice for crossing the Rockies. Hence, the railroad was built across southern Wyoming, climbing west from Cheyenne on "The Gangplank" of Tertiary sediments that lapped onto the Laramie Range, then across the Laramie basin and around the north end of the Medicine Bow Mountains, across the rocky desert of southwest Wyoming and into northeast Utah. Interstate 80 follows much the same route today.

Wyoming boomed during the railroad building days, boomed during the various gold rushes in the mountains, boomed with

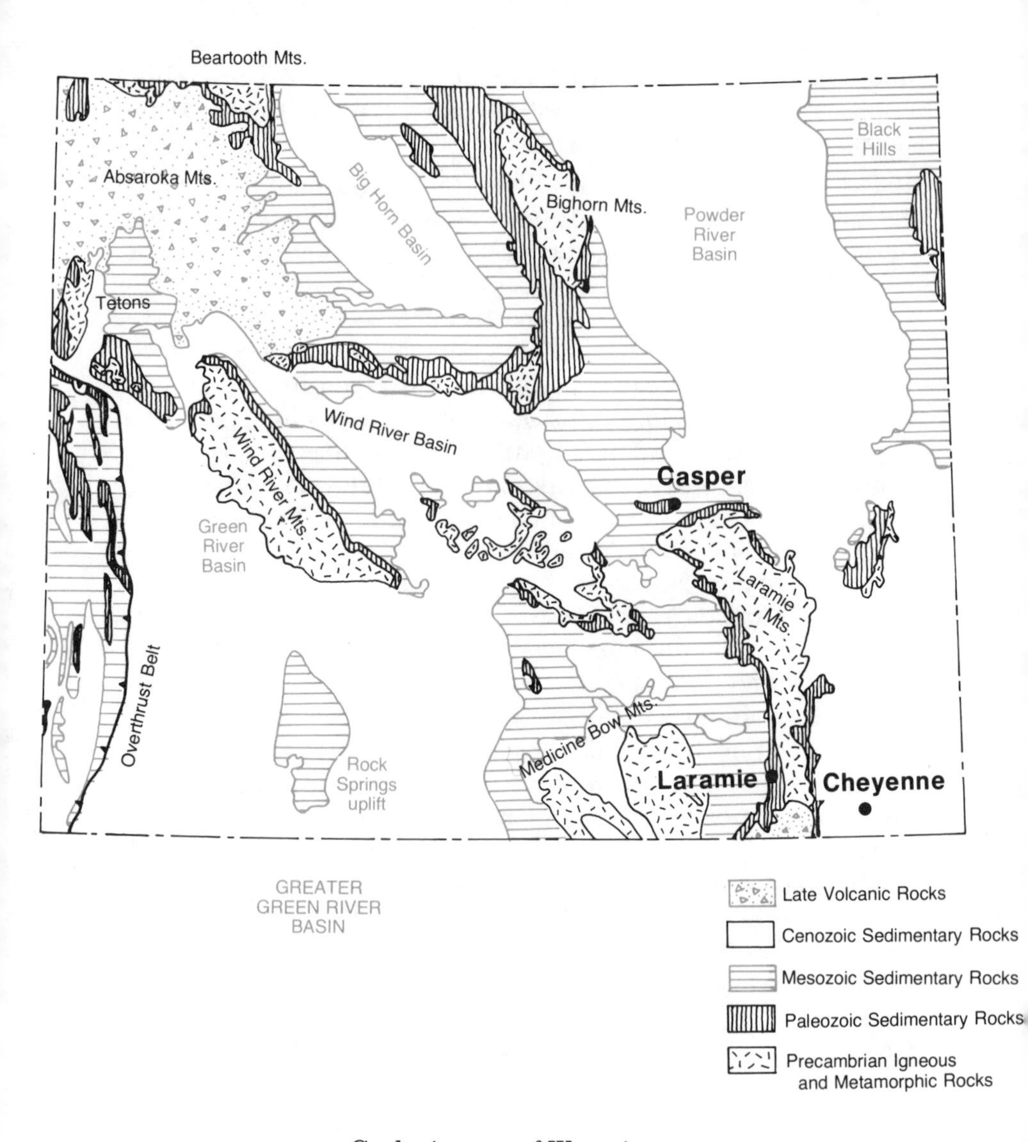

Geologic map of Wyoming.

the tide of oil strikes, boomed with the heyday of uranium mining, and boomed first with underground coal mining and later with surface coal mining. As you can see, the production of natural resources, chiefly energy resources, has been the economic trigger. Wyoming has always had a history of boom followed by bust, followed again by boom. Transient workers came and went with each boom-bust cycle, giving Wyoming the distinctive flavor of a very small permanent population. Indeed, the population today is only about half a million people in a state of 100,000 square miles! There are more antelope than people in Wyoming—that's the way it should be! Wyoming will undoubtedly continue to be driven by boom-and-bust cycles in the future, as the price of natural resources fluctuates on the world market.

Rocks

It is very easy to get "bogged down" in a swamp of technical geological jargon, especially when it comes to names of rocks. Scientists have a tendency to name things, often using lots of complex, Latin-derived words. Our purpose here is to provide a basic introduction to the different kinds of rocks so that you will be equipped to handle the roadside descriptions.

Basically, there are only three groups of rocks that you need to know: 1) igneous rocks, 2) sedimentary rocks, and 3) metamorphic rocks.

Igneous rocks crystallize from magma (molten rock) by cooling. They can either cool within the Earth (intrusive) or on the surface of the Earth (extrusive—lavas and volcanoes). In addition, they are generally classified according to the amount of silica they contain: high-silica igneous rocks are called granite (intrusive) or rhyolite (extrusive); those with intermediate silica are called diorite (intrusive) or andesite (extrusive); and low-silica igneous rocks with about 50 percent silica are called gabbro (intrusive) or basalt (extrusive).

Igneous rocks dominate the landscape in the Absaroka Mountains and Yellowstone National Park of northwestern Wyoming. They also exist in other areas, like the Sweetwater Hills in central Wyoming (where Independence Rock is made of solid granite) and Devils Tower in the northeast corner of Wyoming (the neck of an ancient volcano).

Sedimentary rocks are preserved in Wyoming's large basins, and crop out as tilted layers on the flanks of mountain ranges. Sedimentary rocks are derived from pre-existing rocks through weathering and erosion, followed by deposition in streams, lakes, or the ocean. These rocks form layers and are referred to as "stratified." Sedimentary rocks are generally classified according to the size of fragment that was deposited. From largest to smallest, the size range is: conglomerate (gravel), sandstone (sand-sized), siltstone (silt-sized), and shale (fine mud). In addition, limestone is composed of calcium carbonate derived from shelly organisms (clams, corals, etc.) that lived in clear, warm seas. Lastly, chemical sedimentary rocks are those in arid regions like salt or gypsum that precipitated from water.

Metamorphic rocks are derived from pre-existing rocks through "metamorphosis," caused by extreme heat and pressure. A metamorphic rock has formed new minerals, without melting, that are stable at certain temperatures and pressures. For example, marble is derived from limestone through thermal metamorphism. Likewise, slate is derived from shale through recrystallization at high temperatures and pressures. Schist and gneiss (pronounced "nice") are very common metamorphic rocks, and are found in the cores of Wyoming's mountain ranges. Schist is composed largely of the mineral mica, whereas gneiss is a coarse-grained metamorphic rock with alternating bands of light (quartz and feldspar) and dark (biotite and hornblende) minerals. Schists and gneisses in Wyoming include some of the oldest rocks in North America, over 3 billion years in age!

GEOLOGIC STRUCTURES

Wyoming is a classic region for studying how the crust of the Earth has deformed. The very landscape is the direct result of bending and fracturing of the Earth's crust. We will review the basic types of structures here, and then discuss their origin in the section on geologic history.

Geologic structures may be classified into two basic categories: folds and faults. Folds usually form in relatively ductile (soft) rocks that bend slowly over long periods of geological time, whereas faults occur in more brittle rocks that are rapidly stressed. Therefore, both the relative hardness of the rock and the rate at which it is deformed determine whether a fold or fault will form.

The two most common types of folds are anticlines, in which the limbs dip away from each other, and synclines in which the limbs dip towards each other. The Rawlins uplift and Rock Springs uplift in southwest Wyoming are good examples of anticlines. All of Wyoming's broad, sage-covered basins are synclines. Folds, in general, are the result of compressional (shortening) deformation.

Faults are classified according to whether they extend the crust or shorten the crust. Faults that extend the crust, like the Teton fault, are called normal faults; these faults usually are the result of tension in the crust. Faults that shorten the crust, like the Wind River fault, are called reverse or thrust faults; these are usually the result of crustal compression. A third type of fault, called strike-slip or wrench fault, is the result of shear; the rocks slide past each other. The San Andreas fault in California is a spectacular example of this type of fault. The strike-slip faults scattered around Wyoming are much smaller than the San Andreas!

WYOMING'S EARTH RESOURCES

Wyoming has traditionally been a major producer of energy resources and industrial minerals, and it may play an even bigger role in the future. Wyoming seems to have it all: enormous fields of oil and gas, a seemingly endless supply of low-sulfur coal, large uranium deposits, oil shale, trona, bentonite, gypsum, iron ore, and perhaps most important, clean water.

Wyoming is a leading producer of oil and gas in the Rocky Mountain region. Most is produced from sedimentary rocks buried deeply in the large basins throughout the state, such as the Powder River, Bighorn, Wind River and Green River basins. Refer to the Bighorn Basin chapter for more information on how oil and gas is generated and trapped.

Wyoming contains an incredible amount of coal. Forty thousand square miles, or about 41 percent of the state, is underlain by coal, most of which is close to the surface! This translates into 24 percent of our nation's total coal reserves!

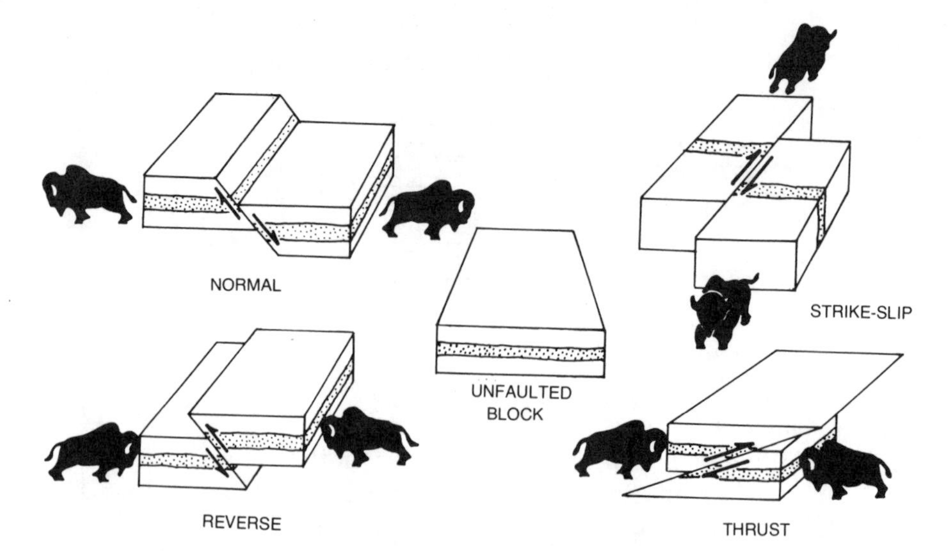

Types of faults: bison push/pulls show how forces act to produce the different types of faults.

Wyoming's coal is in rocks deposited during the Cretaceous period (65 to 140 million years ago) and the early part of the Cenozoic Era (38 to 65 million years ago). During these times, the climate was periodically favorable for dense, swampy vegetation, which formed peat. This peat was then transformed into the almost trillion tons of coal that underlies the sage covered basins of Wyoming.

In the past century, gold was mined to some degree in almost every mountain range in the state, spawning boom towns like South Pass City and Centennial. The town of Encampment sprang up at the base of the Sierra Madre range around the turn of the century as a result of copper mining in that range, and copper mining in the rugged Absaroka Mountains also had its heyday at that time. Although base metal and precious metal production have been historically important, there is no doubt that the production of energy resources has dominated the state's economy.

GEOLOGIC TIME

Time makes geology tick. Not time measured on a watch or in terms of human history, but time measured in millions and billions of years. Time is the essence of geology.

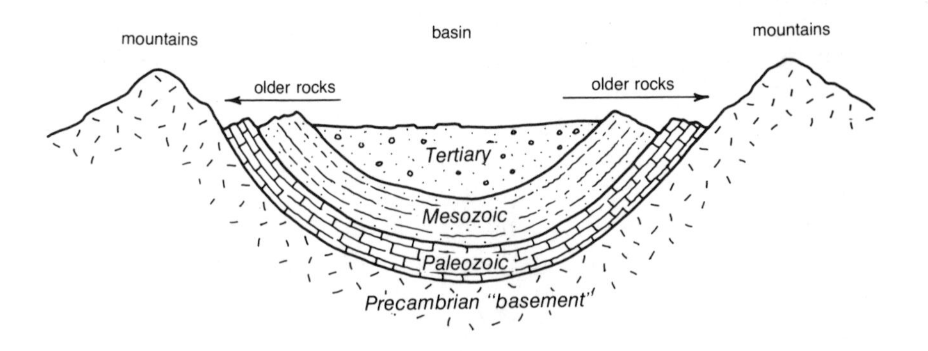

Diagram of stratigraphy in Wyoming's basins and mountain ranges. Note how older rocks are encountered toward the basin edges. This simple diagram shows how the basic structure of Wyoming's foreland province is achieved by upwarping of ranges and downwarping of basins – though in reality more complicated by faults.

Geologists can date rocks and events in Earth history in two fundamental ways: 1) by relative dating techniques, and 2) absolute dating using radioactivity in rocks. Relative dating tells us the relative age of a rock with respect to neighbor rocks, but does not tell us the rock's absolute age in years. For example, if a rock is cut by a vein of quartz, the vein is obviously younger than the rock it cuts—the principle of cross-cutting relations. Also, rock layers at the bottom of a sequence of sedimentary rocks are obviously older than those on top—principle of superposition. However, we must turn to radioactivity to determine the age of a rock in absolute years. Certain igneous and metamorphic rocks contain minerals with radioactive isotopes that decompose at a constant rate. By measuring the amounts of certain isotopes in rocks, geologists can determine the exact age of a rock, plus or minus a small laboratory error.

The geological time calendar is a convenient, man-made scheme for subdividing Earth history based on relative dating, absolute dating, and paleontology. The Earth is estimated to be 4.6 billion years old, although the oldest rocks thus far dated are a little over 4 billion years old (in Australia). The date of 4.6 billion years is based on radioactive dates of meteorites and lunar rocks, which we assume formed at about the same time as the Earth.

All of Earth history, from 4.6 billion years ago to the present, has been subdivided into 4 main "Eras:" 1) the Precambrian, from 4.6 billion years ago to 570 million years ago; 2) the Paleozoic ("ancient life"), from 470 to 230 million years ago; 3) the Mesozoic ("middle life"), from 230 to 65 million years ago; and 4) the Cenozoic ("recent life"), from 65 million years ago to the present. Each of these Eras is, in turn, subdivided into smaller Periods and Epochs. A geologic time scale showing each division, as well as the major events in each, is located on pages viii and ix.

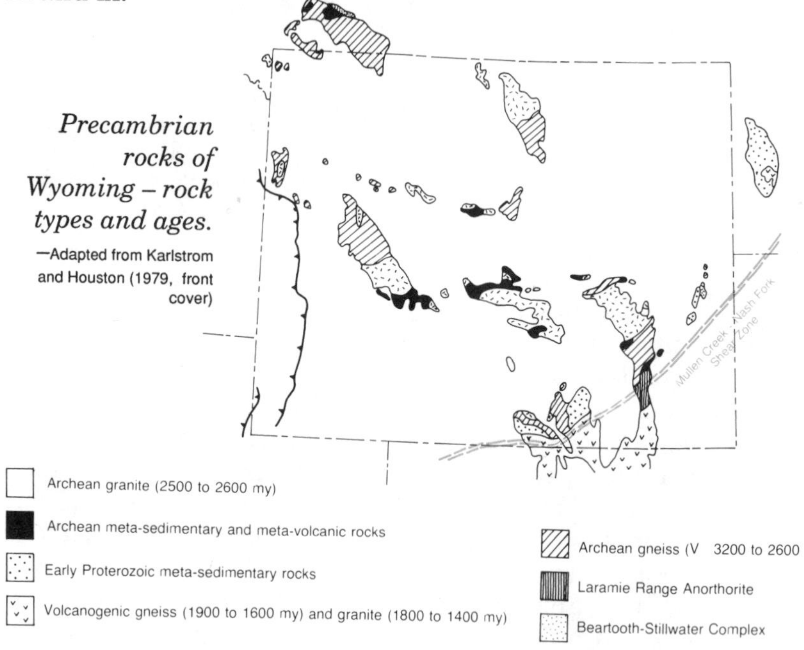

Precambrian rocks of Wyoming – rock types and ages.

—Adapted from Karlstrom and Houston (1979, front cover)

GEOLOGIC HISTORY OF WYOMING

The Beginning—Precambrian Era

The Precambrian history of Wyoming is recorded in the big mountain ranges. Their cores contain schist and gneiss; basement rocks, that are around 3 billion years old, over half the age of the Earth! These are the oldest basement rocks in the western United States. They are referred to as the "Wyoming Province" rocks.

The south margin of the Wyoming Province is faulted against much younger Precambrian basement rocks (1.7 billion years old) that form the high mountains of Colorado, like the Front range and Park range. This fault zone is called the Mullen Creek - Nash Fork fault zone, and it extends northeast through the Sierra Madre and Medicine Bow mountains to the south end of the Black Hills. Because of their antiquity, little is known about the origin of these rocks. These basement terranes probably represent early micro-continents that gathered together during Precambrian time to form the central nucleus of North America, the "craton."

Tropical Seas and Early Life —Paleozoic Era

The Paleozoic history of Wyoming was dominated by warm, shallow, tropical, marine seas. Layers of limestone and shale accumulated on the sea floor to build the Paleozoic sedimentary sequence.

The Rocky Mountain states were part of a broad, submerged, shallow continental shelf that extended along the western margin of North America to the Yukon. Wyoming's portion of the shelf was near the Equator. Land was exposed only in the central part of the North American continent. Marine waters periodically transgressd over the western shelf, depositing sandstone, shale and limestone with each incursion of the sea. The Cambrian Gros Ventre formation of limestone and shale,

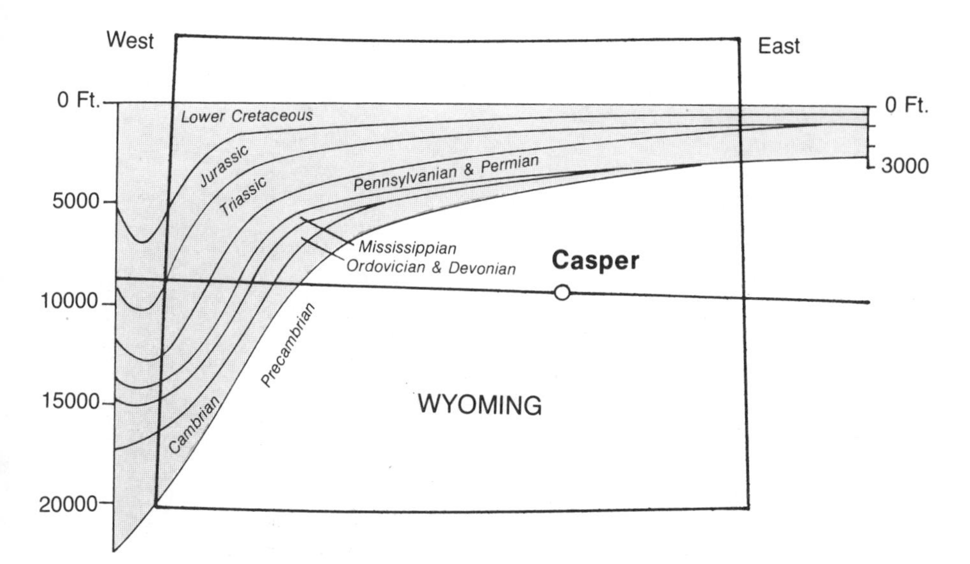

The thickness of the sedimentary rock pile in Wyoming increases dramatically to the west. A shallow platform in the east and an ever-deepening basin to the west existed throughout most of Paleozoic time.

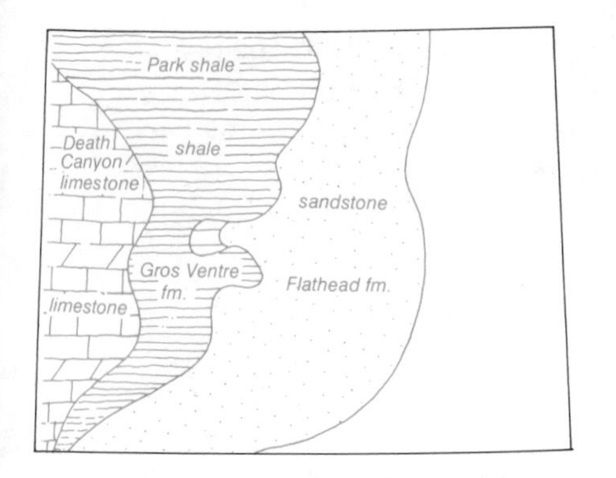

Cambrian rock types and formations in Wyoming.

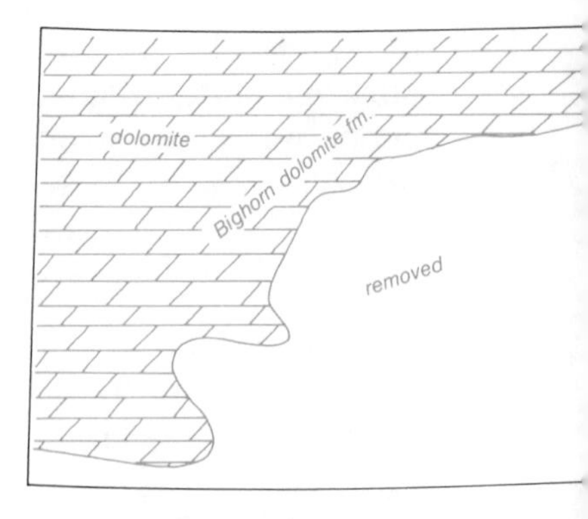

Ordovician rock types and formations in Wyoming.

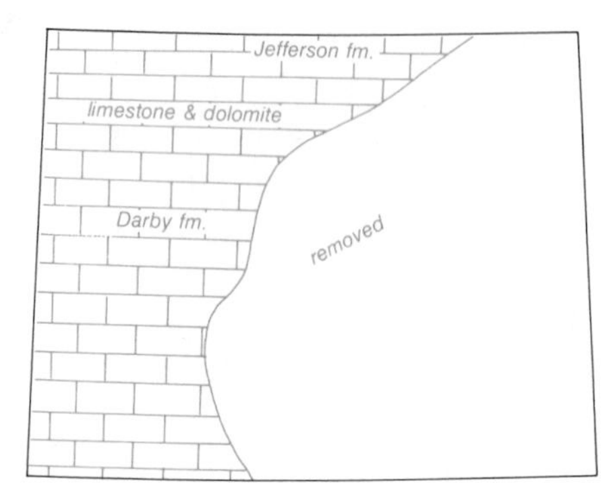

Devonian rock types and formations in Wyoming.

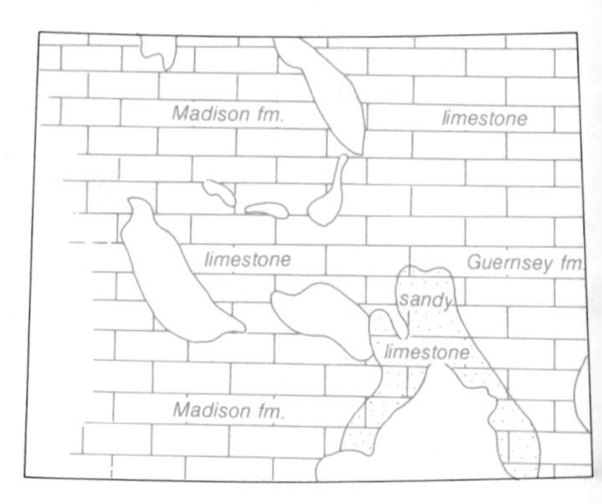

Mississippian rock types and formations in Wyoming.

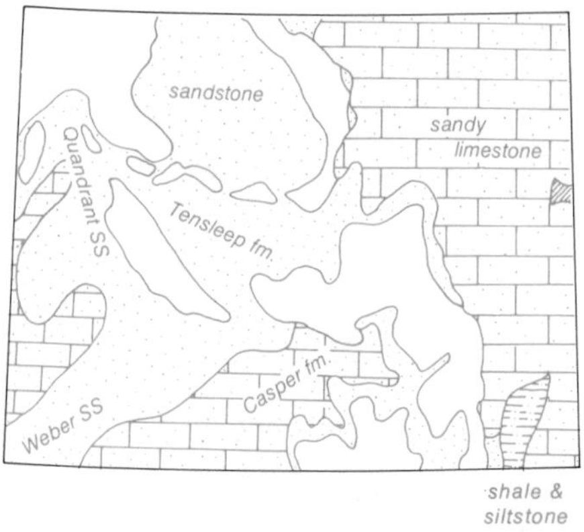

Pennsylvanian rock types and formations in Wyoming.

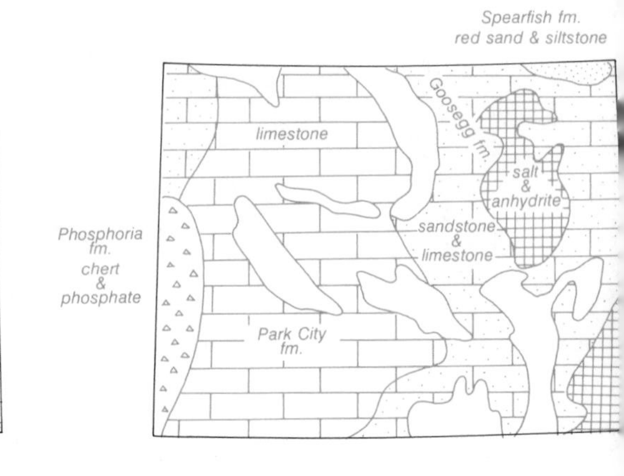

Permian rock types and formations in Wyoming.

–Adapted from R.M.A.G. *Geologic Altas* (1972)

the Ordovician Bighorn dolomite, and the famous Mississippian Madison limestone are examples of the marine sediments deposited on the Wyoming shelf during Paleozoic time. The Wyoming shelf was basically stable during this time, although minor up-and-down movement of the crust caused either erosion or nondeposition of sediments, producing unconformities in the stratigraphic record.

In Pennsylvanian time, the ancestral Rocky Mountains of Colorado were uplifted, perhaps in response to the collision of North America and North Africa. A northern prong of the ancestral Rockies extended into southeast Wyoming from Colorado, called the Pathfinder uplift, but basically the Wyoming shelf was the site of sand deposition (Tensleep sandstone) during the Pennsylvanian.

On a global scale, the close of Paleozoic time witnessed the collision and accretion of all the world's continental land masses to form a supercontinent called Pangaea. The northern part of Pangaea, Laurasia, was made of North America and Europe; the south part, Gondwanaland, was made of South America, Africa, Antarctica, India and Australia. Since this time, the continents have moved to their present positions by a process called "plate tectonics."

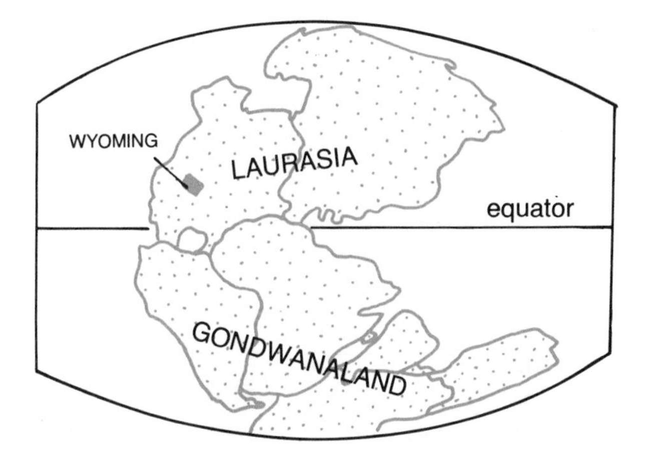

Continental reconstruction at the end of the Paleozic Era, showing the Supercontinent of Pangaea, composed of Laurasia and Gondwanaland.

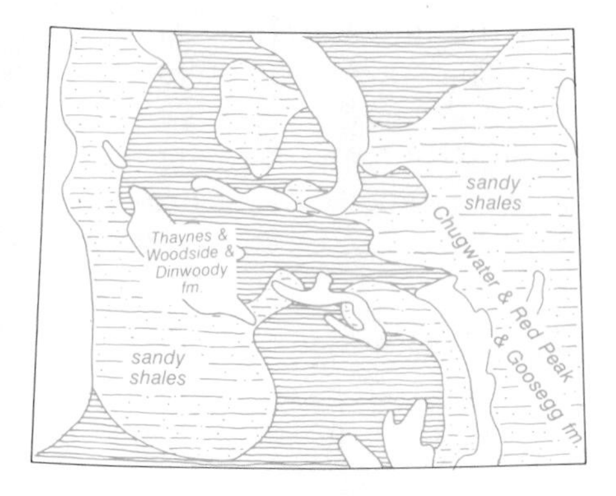

Triassic rock types and formations. Most are bright red rocks in Wyoming.

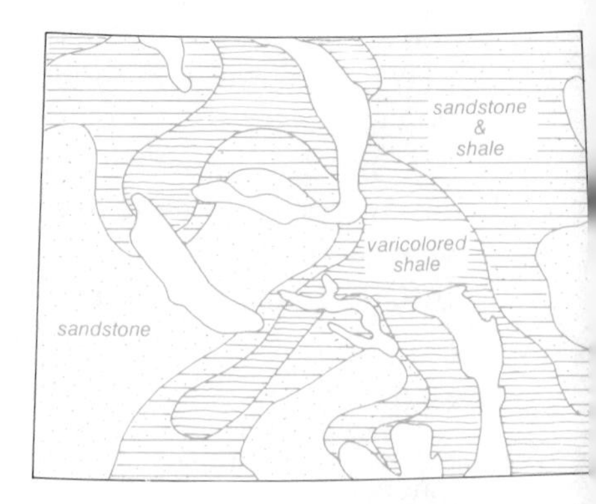

Jurassic rocks of the Morrison formation in Wyoming.

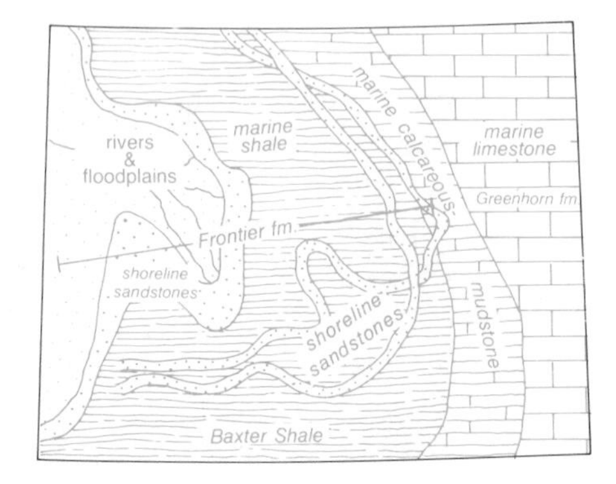

–Adapted from R.M.A.G. *Geologic Altas* (1972)

Cretaceous rock types and formations in Wyoming.

Redbeds and Dinosaurs —Mesozoic Era

Mesozoic time was a period of transition for the Rocky Mountain region. Bright red Triassic sandstones and shales, the Chugwater formation, were deposited across Wyoming, in sharp contrast to the gray Paleozoic marine limestones. The red and green, dinosaur-bearing mudstones of the Jurassic Morrison formation were deposited on floodplains of rivers, again in contrast to earlier marine shelf sediments.

Marine conditions returned to Wyoming during Cretaceous time, but this time as an inland seaway that extended from the Gulf Coast to the Arctic Ocean, and bordered to the west and east by land. The Rocky Mountain states were largely covered by this seaway, and received thick deposits of sandstones and black, organic-rich shale, such as the Thermopolis, Mowry and Cody shales. These black shales have been the source rocks for much of the region's oil.

Mountain Building and Mammals —Cenozoic Era

Cenozoic time was a period of mountain building and continental sedimentation. Marine waters never returned to the Rocky Mountain region after Cretaceous time.

The Rocky Mountains were uplifted during an event called the Laramide orogeny—orogeny means mountain building. This event started in late Cretaceous time and continued into early Eocene time. The Rocky Mountain region was compressed as a result of subduction of oceanic crust along the western

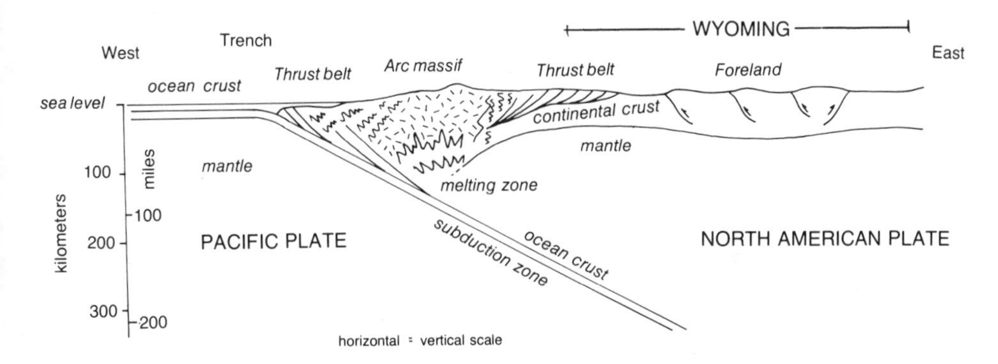

Tectonic framework of western North America during the Laramide orogeny. –Adapted from Grose (1972, p.44)

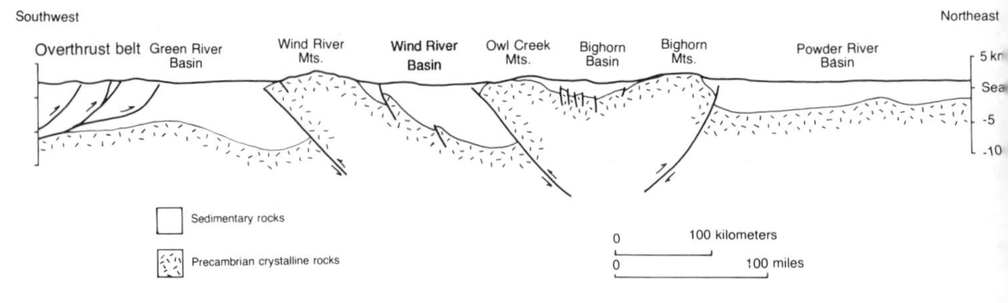

A profile from northeast to southwest across Wyoming.

margin of the continent. This same process is occurring today along the west coast of South America, and is producing a mountain belt on the east flank of the Andes volcanic chain that is very similar to the Laramide Rocky Mountains of North America. When the crust is squeezed or compressed, it shortens by faulting along reverse or thrust faults. Laramide thrust faults abound in Wyoming and define many of the Precambrian-cored uplifts like the Wind River Range, Bighorn Range, and Laramie Range. Because the sedimentary rocks above the Precambrian basement are more ductile, they were folded over the top and sides of the thrusted basement blocks. This style of deformation, sometimes called foreland style, is spectacularly demonstrated at the Clarks Fork Canyon on the south end of the Beartooth Range near Cody, Shell Canyon in the western Bighorn Range near Greybull, and along the southwest flank of the Gros Ventre Range south of Jackson.

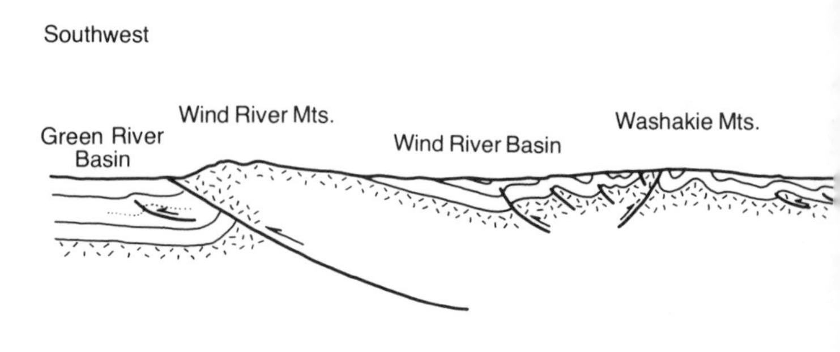

Cross section of Wyoming from

Wyoming yielded to compressive stresses in another way. The far western part of the state is called the overthrust belt for the distinctive style of deformation displayed there. The overthrust belt is a more or less continuous zone of thrust faults and folds that extend along the "backbone" of North America from the Brooks range in Alaska, through the Rocky Mountains of Canada, and down through western Montana, western Wyoming and central Utah. The overthrust belt is characterized by an array of west-dipping, low-angle thrust faults and associated folds that, in general, do not directly involve crustal basement rocks. In other words, the thrusted sheets of rock lay like shingles on a roof, and the thrust faults do not cut into the roof. This region became a major oil producing province in southwest Wyoming and adjacent Utah in the mid-1970s when enormous amounts of oil and gas were found in faulted anticlines.

Highlands created by the Laramide orogeny were slowly worn down by weathering and erosion, with the eroded debris filling adjacent basins. Paleocene time was more humid and wet than our present climate and thick, swampy deciduous forests grew across the state. Their accumulated debris formed the great coal beds of the Fort Union formation; some of those beds are more than 100 feet thick!

Eocene time is most noted for the accumulation of oil shale beds in southwest Wyoming and adjacent parts of Utah and Colorado. These organic-rich shales, called the Green River formation, accumulated in large, shallow, playa lakes called Lake Gosiute and Lake Uinta. Countless fish skeletons, delicately preserved in the laminated shales of the Green River formation, can be seen at Fossil Butte National Monument west of Kemmerer, Wyoming.

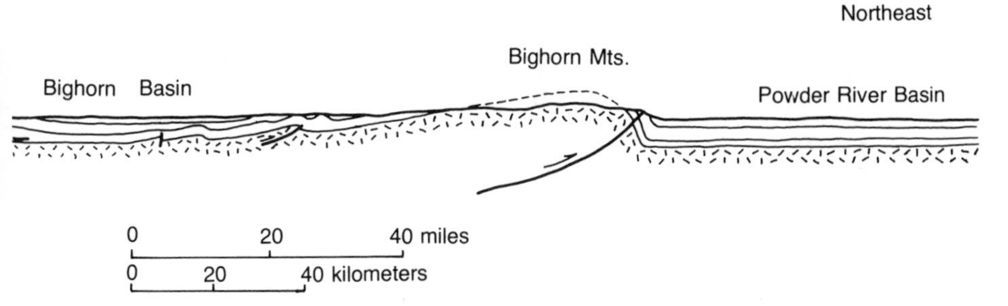

Green River Basin to Powder River Basin

Wyoming sits on the very eastern edge of the North American Thrust Belt.

Oligocene, Miocene and Pliocene time saw continued erosion in the mountains and deposition of the sediments in the basins. Large volcanic eruptions of rhyolite ash occurred to the west in the Basin and Range province of Nevada and Utah during this time, and the ash was carried east by the prevailing winds. This ash is largely responsible for the stark white color of the Oligocene Wind River formation as seen, for example, in the Shirley Basin south of Casper. Miocene and Pliocene rocks have been uniquely preserved in the Sweetwater Hills (also called Granite Mountains) in central Wyoming; this area has been down-faulted since Pliocene time, thus preserving these young Cenozoic strata.

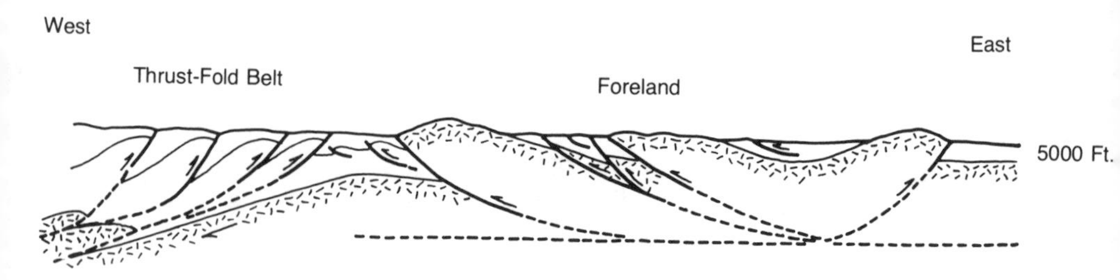

Cross section of Wyoming Thrust and Foreland structure. –Adapted from Lowell (1983, p. 2)

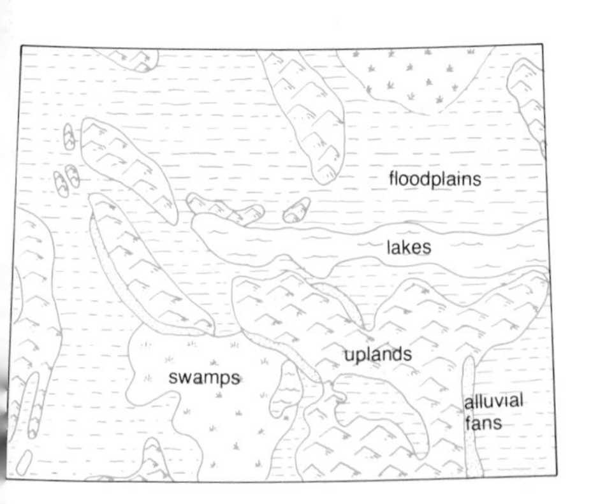

Paleocene rocks and formations

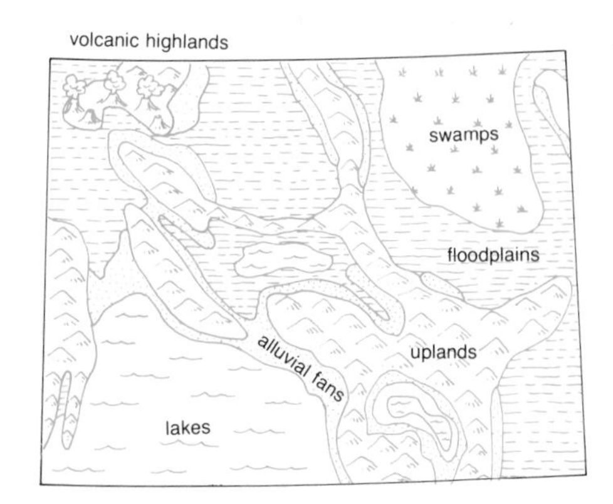

Eocene facies and formations

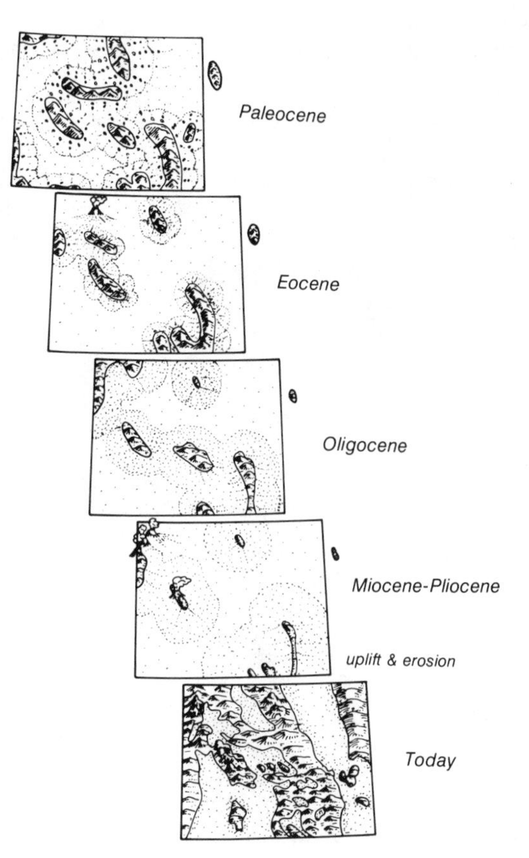

A sequence of time maps of Wyoming showing the history of "fill-up" of the state's basins throughout the Tertiary. Uplift in Miocene-Pliocene time caused stream cutting and exposure of once-buried mountain ranges, as seen today. –Adapted from R.M.A.G. *Geologic Altas* (1972)

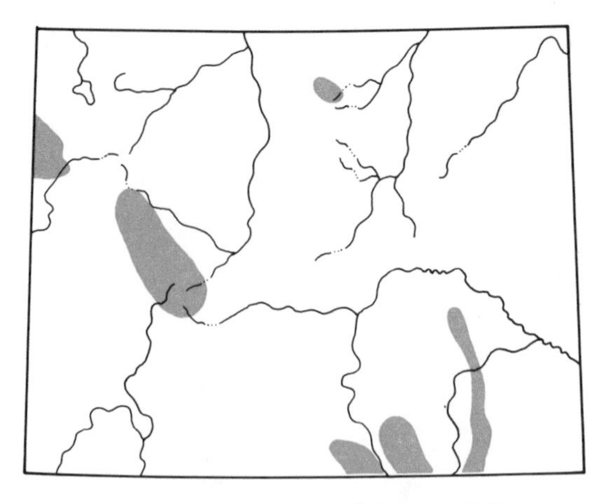

Present-day streams superimposed on Miocene-Pliocene topography. The predecessors to today's streams probably did not vary too much from today's stream locations. –Adapted from R.M.A.G. *Geologic Altas* (1972)

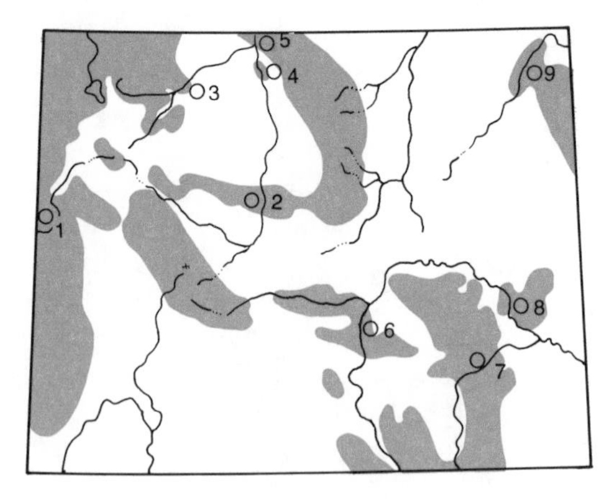

1 - Snake River Canyon—through the Thrust Belt Ranges
2 - Wind River Canyon—cuts across Owl Creek Mountains
3 - Shoshone River Canyon—slices Rattlesnake/Cedar Mountains
4 - Bighorn River—cuts Sheep Mountain
5 - Bighorn River Canyon—goes through Bighorn Mountains
6 - Fremont Canyon—North Platte River cuts through Seminoe Mountains
7 - Laramie River—flows through Laramie Range
8 - Platte River—flows through canyon at Guernsey State Park across the Hartsville uplift
9 - Belle Fourche River—cuts across the structure at the north end of the Black Hills

After Miocene-Pliocene uplift, streams began to cut down and erode. As they did so, they cut into buried mountain ranges and entrapped themselves into canyons we see today. The circles show where rivers cut right across ranges in patterns that seem senseless until this story of sediment fill-up, uplift, erosion and entrenchment is understood. –Adapted from R.M.A.G. *Geologic Atlas* (1972)

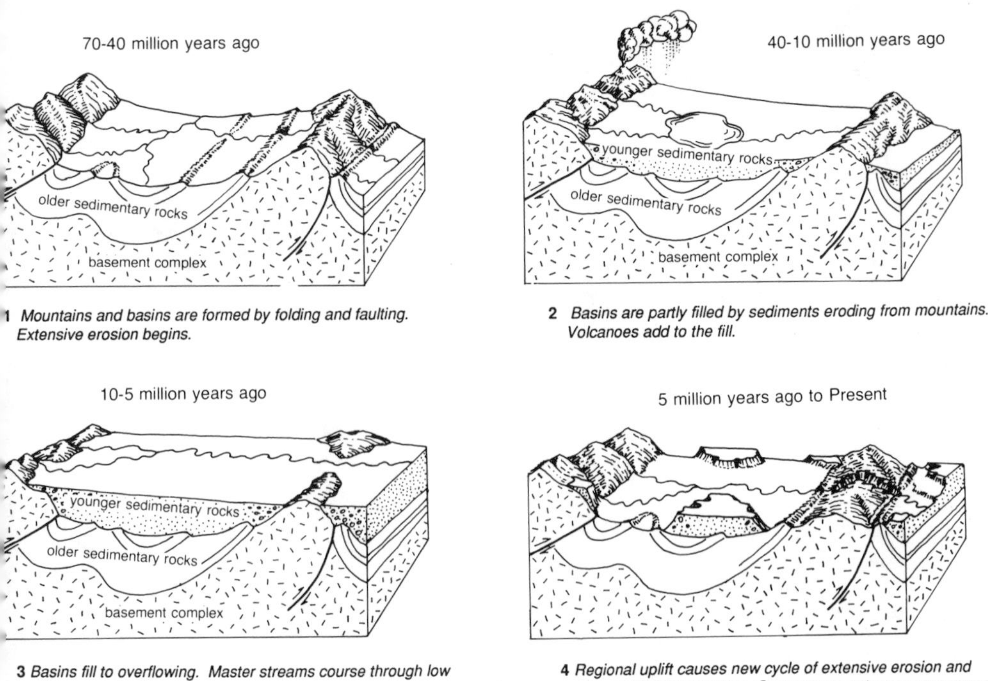

1 *Mountains and basins are formed by folding and faulting. Extensive erosion begins.*

2 *Basins are partly filled by sediments eroding from mountains. Volcanoes add to the fill.*

3 *Basins fill to overflowing. Master streams course through low places over buried ranges.*

4 *Regional uplift causes new cycle of extensive erosion and downcutting by streams. Canyons are cut across ranges as mountains are exhumed.*

–Adapted from a drawing by S.H. Knight, University of Wyoming

In the last 2 million years, Quaternary time, several dramatic things have significantly changed the Wyoming landscape. First, Yellowstone came into existence through several, enormous volcanic explosions of rhyolite ash apparently due to eastward movement of the Snake River plain; this is discussed in the section on Yellowstone Park. Secondly, the Teton Range rose as Jackson Hole dropped, both movements the result of tensional stresses that are pulling the crust of the western U.S. apart. Lastly, on a far grander scale than either Yellowstone or the Tetons, the entire intermountain region and Basin and Range province have been arched upwards causing rivers to down-cut their channels, resulting in accelerated erosion over the entire western U.S. The cause of this uplift may be related to high heat flow from the mantle which is causing crustal extension and normal faulting. In Wyoming, the result is that the old Laramide ranges and basins are being exhumed, or dug out, as modern rivers cut their way downward. This is why you will see rivers like the Wind River cutting across the Owl Creek and Bighorn mountains, instead of flowing around them. John Wesley Powell, early explorer and western geologist, called this process "superposition" of streams.

In conclusion, you can see that Wyoming has had a long and complex geological history. The story has been greatly simplified for our purpose here, but it is nevertheless based entirely on what the rocks tell us. Our ideas and interpretations may change with time as more is learned, but the basic facts recorded in the rocks remain true. As geologists, we are constantly striving to better understand the history of the Earth as recorded in the rocks.

I
Southeastern Wyoming

INTRODUCTION

Southeastern Wyoming is a zone of transition from the High Plains to the Rocky Mountains. It is a land that contains elements of both the plains and mountains, for here the north end of the rugged Colorado Rockies project into an otherwise flat region. For more than 150 years, herds of pronghorn antelope have curiously watched a steady caravan of passing humanity. Southeastern Wyoming was the corridor through the Rockies for early explorers, fur trappers, west-bound emigrants, and the Union Pacific Railroad. Today, it remains a thoroughfare for millions of travellers who, like the early pioneers, pass through but seldom stop.

To the south, the northern Colorado Rockies are composed of two great mountain uplifts, the Front Range on the east and the Park Range on the west. The Front Range rapidly looses elevation northward from the 14,000 foot summit of Longs Peak, Colorado, and splits into the Laramie and Medicine Bow ranges in southeastern Wyoming. Similarly, the Park Range plunges north into Wyoming, becoming the Sierra Madre Range. The landscape of southeastern Wyoming is dominated by these three prongs or extensions of the Colorado Rockies. In addition, a small arch called the Hartville uplift connects the northern Laramie Range with the south end of the Black Hills in South Dakota.

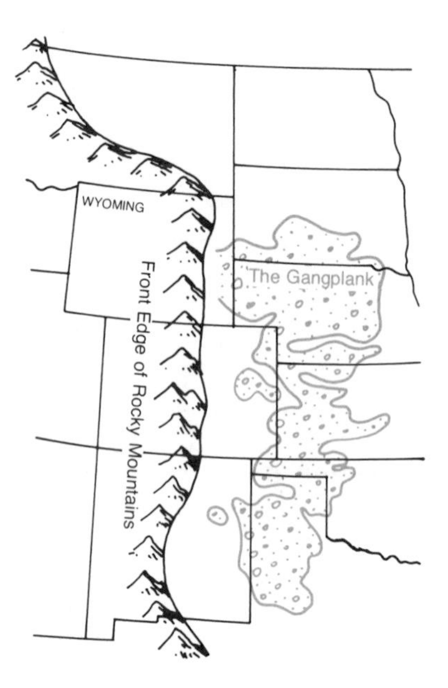

Extent of the high plains east of the Rocky Mountain front. The Gangplank is part of a remnant apron of early Tertiary-aged sediments that spread east from the eroding Rocky Mountains.

MOUNTAINS OF SOUTHEASTERN WYOMING

Laramie Range

The Laramie Range is a mere prong of the northern Colorado Front Range, but it is this range, out of all others in the Rockies, that gave the great mountain building episode called the Laramide orogeny its name. The Laramide orogeny was the birth of the entire Rocky Mountain system, occurring in early Cenozoic time, around 50-65 million years ago. It was caused by subduction of oceanic crust along the western margin of the continent; this in turn compressed and uplifted the continental crust to form the Rockies.

By Rocky Mountain standards, the Laramie Range is a low, flat, subdued uplift. It barely resembles the mass and height of its Colorado cousin to the south. Because the Laramie Range was not so formidable, western emigration moved through this part of the country in the 1800s. The highest point is Laramie Peak, 10,274 feet, at the north end of the range. Laramie Peak is visible for more than 100 miles and was the pioneer's first glimpse of the Rocky Mountains. The emigrants followed the North Platte River, skirting the north end of the range to Fort Caspar. Later, in the late 1860s, the Union Pacific Railroad

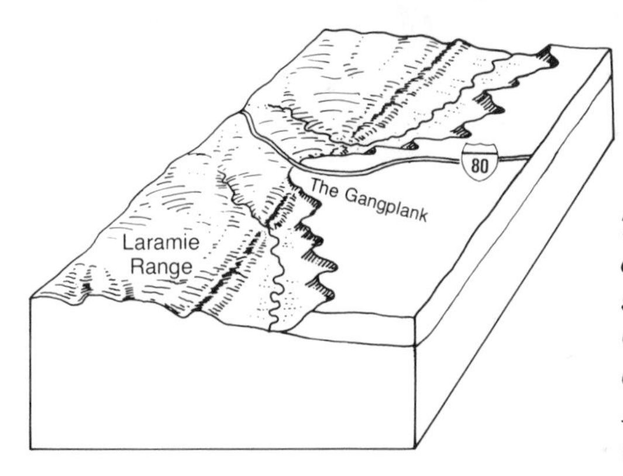

The Gangplank, of partially eroded Tertiary-aged sedimentary rocks west of Cheyenne, allows an easy road and railroad climb over the Laramie Range. –Adapted from Blackstone (1988)

was constructed across the south part of the range between Cheyenne and Laramie on what is known as the Gangplank, a gently sloping surface formed by young Miocene sediments that lap onto the southern Laramie Range, and provide a ramp across it. This route was shorter than the Oregon Trail.

The Laramie Range may be subdivided into northern and southern halves according to the type and age of bedrock. The northern half of the range, from Casper to Wyoming 34 south of Wheatland, is composed of extremely ancient granitic rocks of Precambrian age. These rocks were intruded as hot magma into the primitive crust of Wyoming around 2.6 billion years ago. South of Wyoming 34, the Laramie Range is composed of much younger Precambrian granitic rocks, called the Sherman

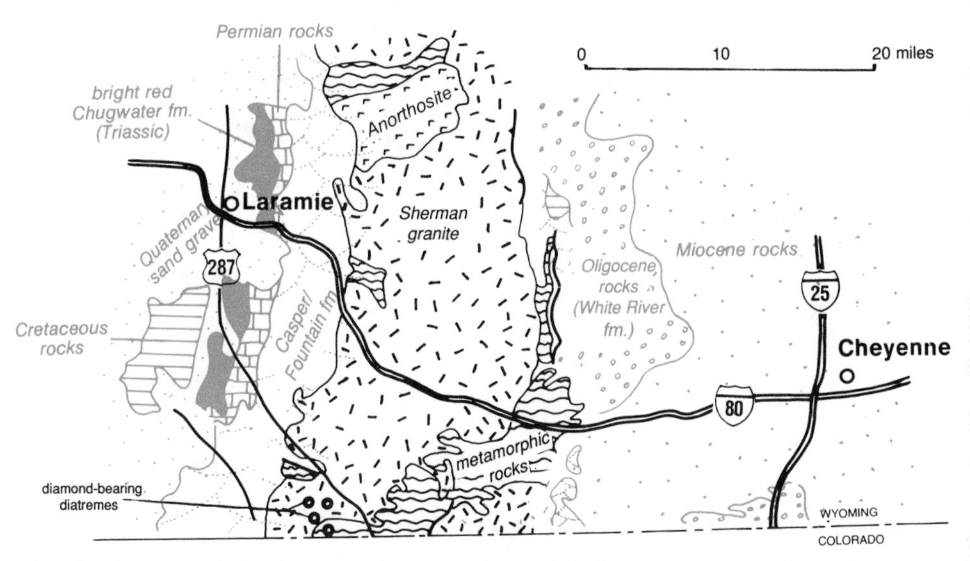

Geologic map of the Laramie Range along I-80 between Laramie and Cheyenne.

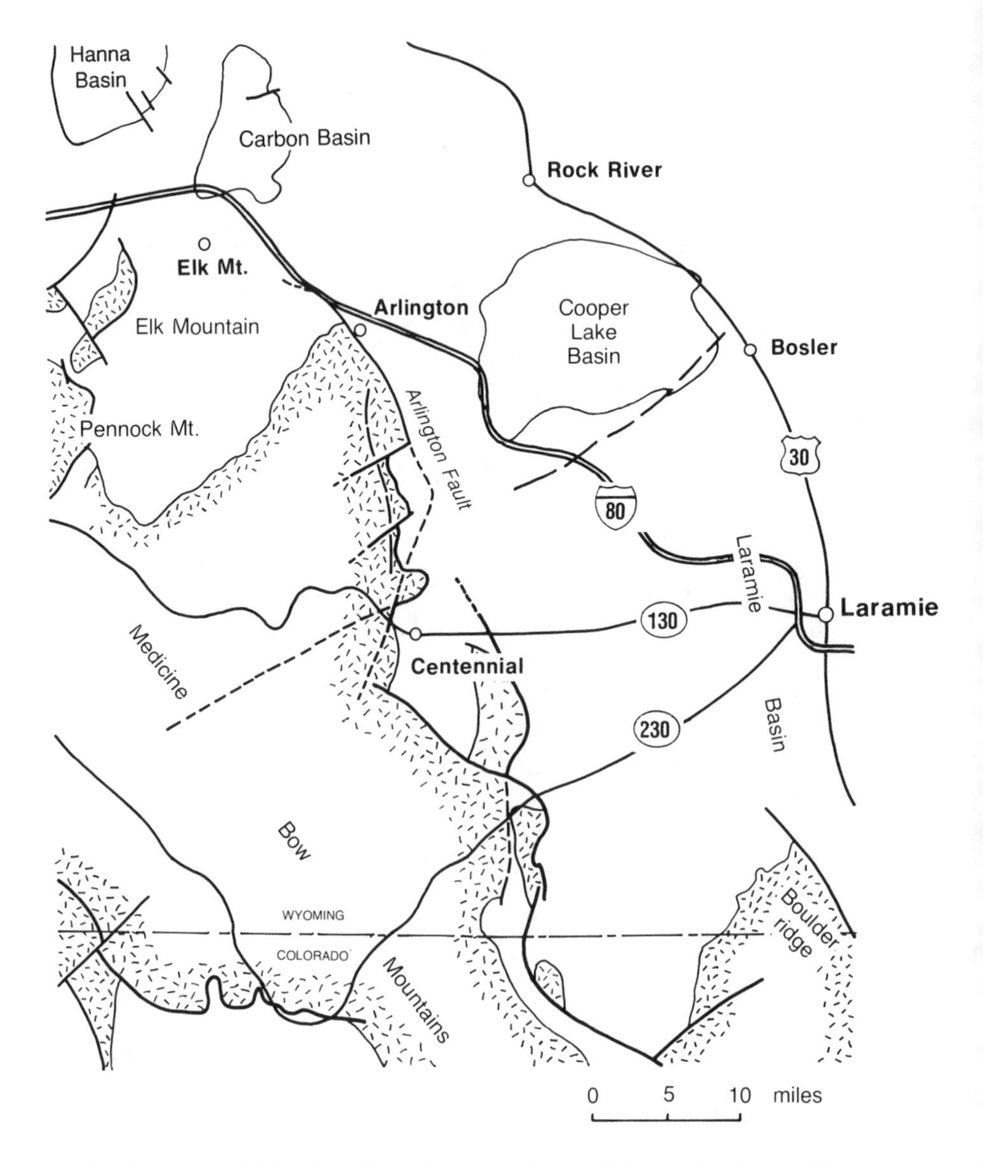

Index map of Medicine Bow Mountains and Laramie Basin. Note sedimentary rocks (white), Precambrian "basement" rocks (hachured), and faults (dark lines). –Adapted from Blackstone (1980)

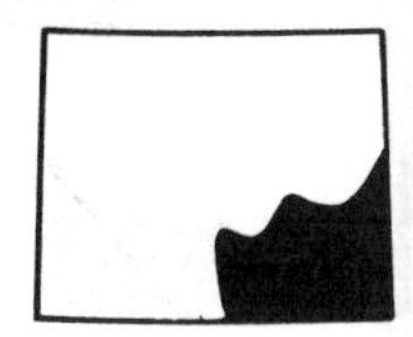

granite, that crystallized around 1.4 billion years ago. The Sherman granite in turn contains a unique body of igneous rock called anorthosite, composed almost entirely of the mineral plagioclase feldspar.

Seismic reflection studies have shown that the eastern flank of the range is bounded by a major thrust fault that dips back under the range. This shows the Laramie Range was uplifted and shoved relatively to the east over the plains during the Laramide orogeny. Similarly, Casper Mountain at the north end of the range was thrust northward over the margin of the Powder River Basin.

Medicine Bow Mountains (Snowy Range)

The Medicine Bow Mountains, sometimes called the "Snowy Range" for the sparkling, sugar-white outcrops of quartzite along its crest, are the west prong of the Colorado Front Range. The name is of Indian derivation: Indians came here to collect ash wood to make their bows and arrows for hunting.

Like the Laramie Range to the east, the Medicine Bow Range may be subdivided into northern and southern halves. The north half contains granite and gneiss that are older than 2.6 billion years, overlain by metamorphosed sedimentary rocks, quartzite and schist, that are around 2 billion years old. The crest of the range is composed of these quartzites that have been turned on end to form spectacular cliffs. The southern part of the range contains much younger granite that is around 1.4 billion years old. To keep this in perspective, there is a billion years difference in the age of bedrock from the north to the south ends of the range!

Also, like the Laramie Range, the Medicine Bows were raised and shoved to the east during the Laramide orogeny by a thrust fault along the east flank of the range. This fault, called the Arlington thrust fault, is well exposed north of Wyoming 130 near Centennial, where dark-colored Precambrian rocks lie on much younger Cretaceous sedimentary rocks.

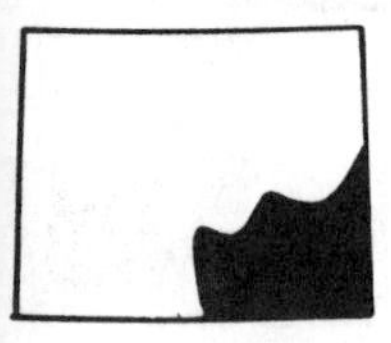

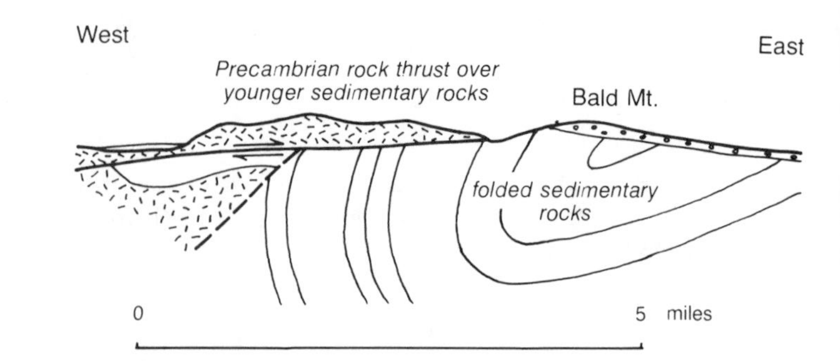

Cross section of Arlington fault just north of Centennial, Wyoming.
–Adapted from Blackstone (1980)

Sierra Madre Mountains

The Sierra Madre Mountains are the north end of the Colorado Park Range, separated from the Medicine Bows by the narrow Saratoga Valley. The Washakie Basin extends west of the range.

Copper was discovered in the Sierra Madres in 1896, leading to the construction of numerous mines and small towns. From 1899 to 1908, 24,000 pounds of copper were produced, principally from the famous Ferris-Haggarty Mine. An aerial tramway 16 miles long was constructed to carry copper ore from the mine to a smelter at Encampment. High-grade ore from the mine ran from 30 to 40 percent copper, with some silver and gold. The Ferris-Haggarty Mine closed in 1908 after the company was indicted for fraudulent stock sales, but ore deposits remain in the area and mining potential still exists. A museum at Encampment features the geology and mining history of the area.

Like the Laramie and Medicine Bow ranges, the Sierra Madre has the same pattern of Precambrian bedrock ages, with rocks older then 2.5 billion years in the north half, and less than 2.5 billion in the south half. Like other ranges in Wyoming, these ancient rocks rose during the Laramide orogeny, 50 to 65 million years ago.

MULLEN CREEK—NASH FORK SHEAR ZONE

If you read the summaries of the three ranges in southeastern Wyoming, you probably noticed a repetitive theme: the

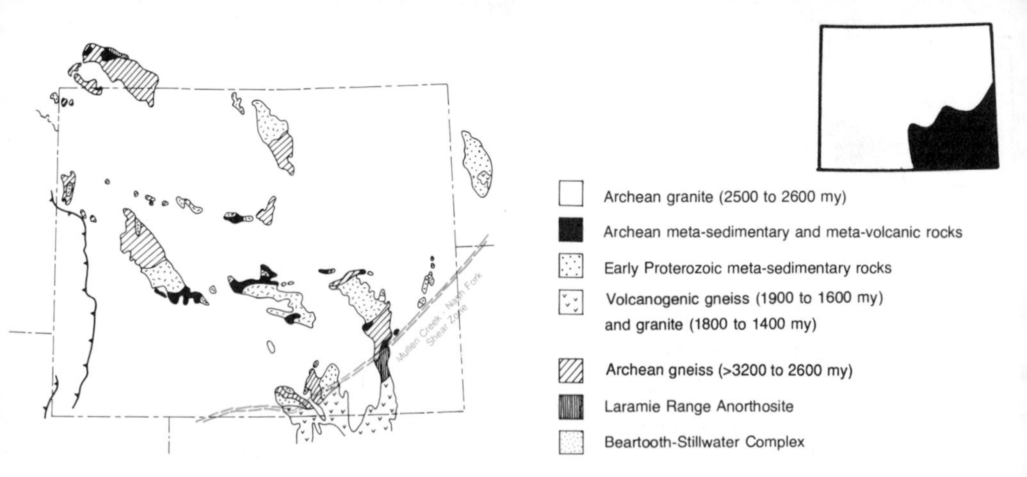

Precambrian rocks of Wyoming – rock types and ages. —Adapted from Karlstrom and Houston (1979, front cover)

dramatic difference in the ages of Precambrian rocks in all three ranges. Most mountains in Wyoming, like the Wind River, Bighorn, and Teton ranges, contain granite, gneiss, and schist that are around 3 billion years in age, give or take a few hundred million. But in southeastern Wyoming, we find much younger Precambrian rocks at the south ends of the ranges. Why is this so?

The different ages may be explained by the large fault that cuts through southeastern Wyoming. This fault, called the Mullen Creek—Nash Fork shear zone, trends northeast through the Sierra Madre, Medicine Bow, and Laramie ranges, then parallels the Hartville uplift to the south end of the Black Hills. It is the boundary between two pieces of continental crust that joined between 1.6 and 1.7 billion years ago. Although this may seem fantastic, remember that 1.7 billion years ago was a very long time ago indeed, and North America did not look anything like it does today. The Rockies and Appalachians did not exist and the center of the continent was slowly being assembled through the accretion of smaller chunks of primitive granitic crust. The Mullen Creek—Nash Fork shear zone is the boundary between two of these chunks.

SEDIMENT-FILLED BASINS OF SOUTHEASTERN WYOMING

Numerous sediment-filled, grass-covered basins are scattered through southeastern Wyoming. These were home to

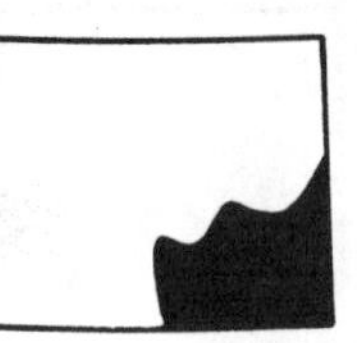

vast buffalo herds, and today support countless pronghorn antelope. It's been said that there are more antelope than people in Wyoming!

The north end of the Denver Basin lies east of the Laramie Range and south of the Hartville uplift. The Denver Basin is a large trough along the base of the Colorado Front Range. It contains older Paleozoic and Mesozoic marine strata, as well as younger sediments eroded from the Front Range.

The Laramie, Hanna, and Shirley basins are west of the Laramie Range. The Laramie Basin lies between the Laramie and Medicine Bow ranges and extends north to Como Bluffs anticline, the site of a famous dinosaur quarry. The Hanna Basin is an extremely deep, funnel-shaped depression between the Medicine Bow Mountains and the southeastern corner of the Granite Mountains. It has the distinction of being the deepest and smallest basin in the Rocky Mountains, filled with over 40,000 feet of sedimentary rock, most of it Cenozoic in age. The Hanna Basin has been a prolific coal producer from Paleocene sedimentary rocks. The Shirley Basin, on the northwest side of the Laramie Range, is covered with young sedimentary rocks that contain important deposits of uranium.

DINOSAURS AT COMO BLUFF

Como Bluff was the site of an intriguing feud between two dinosaur-digging paleontologists in the last century. But first, let's understand the geology. Como Bluff is one of two prominent anticlines, or up-folds, at the north end of the Laramie Basin near Medicine Bow. The bluffs are two resistant layers of lower Cretaceous sandstone that dip southward on the south limb of Como Bluff anticline. On the north side of Como Bluff, the Jurassic Morrison formation is exposed near the center of the fold. This formation has produced many large dinosaur fossils from other Rocky Mountain sites, including Dinosaur National Monument in Colorado. The Morrison formation contains shales and sandstones deposited by rivers about 180 million years ago, during the heyday of the dinosaurs. When

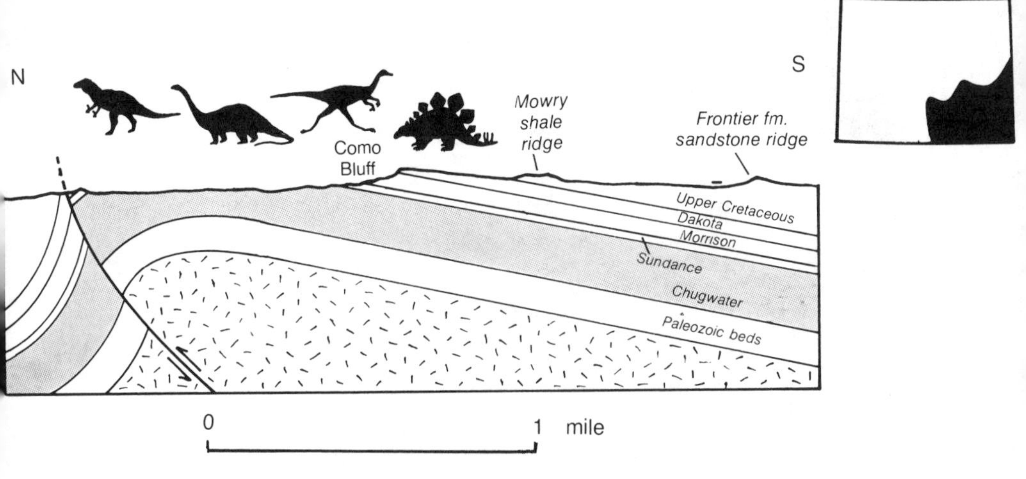

Cross section of Como Bluff. –Adapted from Blackstone (1988)

dinosaurs died, they were rapidly buried by river sediments and their bones eventually fossilized by minerals precipitated from the ground water. During the Laramide orogeny, these anticlines were heaved up and eventually eroded to expose the Jurassic Morrison formation and its fossil treasures.

The paleontologist Othniel Charles Marsh, from Yale's Peabody Museum, was directed to Como Bluff in 1877 by two local railroad men. Marsh and his assistants opened several quarries over the next two years, which yielded tons of well-preserved dinosaur skeletons, including Brontosaurus. Meanwhile, Edward Drinker Cope, a wealthy Philadelphian who was digging dinosaurs near Canyon City, Colorado, could not resist the news of "huge flesh-eating saurians" found in Wyoming and "invaded" the quarries with his own crew. After several violent encounters with the Marsh crew, Cope established his own quarry at Como Bluff in 1879. The two camps frequently fought and sabotaged each other, and soon the entire continent west of the Mississippi was not big enough for both Cope and Marsh. Although both men went on to explore other areas, they remained fierce competitors till death, each trying to out-do the other with new discoveries and voluminous publications.

Today, the Como Bluff cabin by US 30, which is constructed entirely of dinosaur bones, stands as a monument to this rich deposit of Jurassic dinosaurs. No fossil bones remain on the surface at Como Bluff, but complete dinosaur skeletons from here are on display in major museums throughout the world.

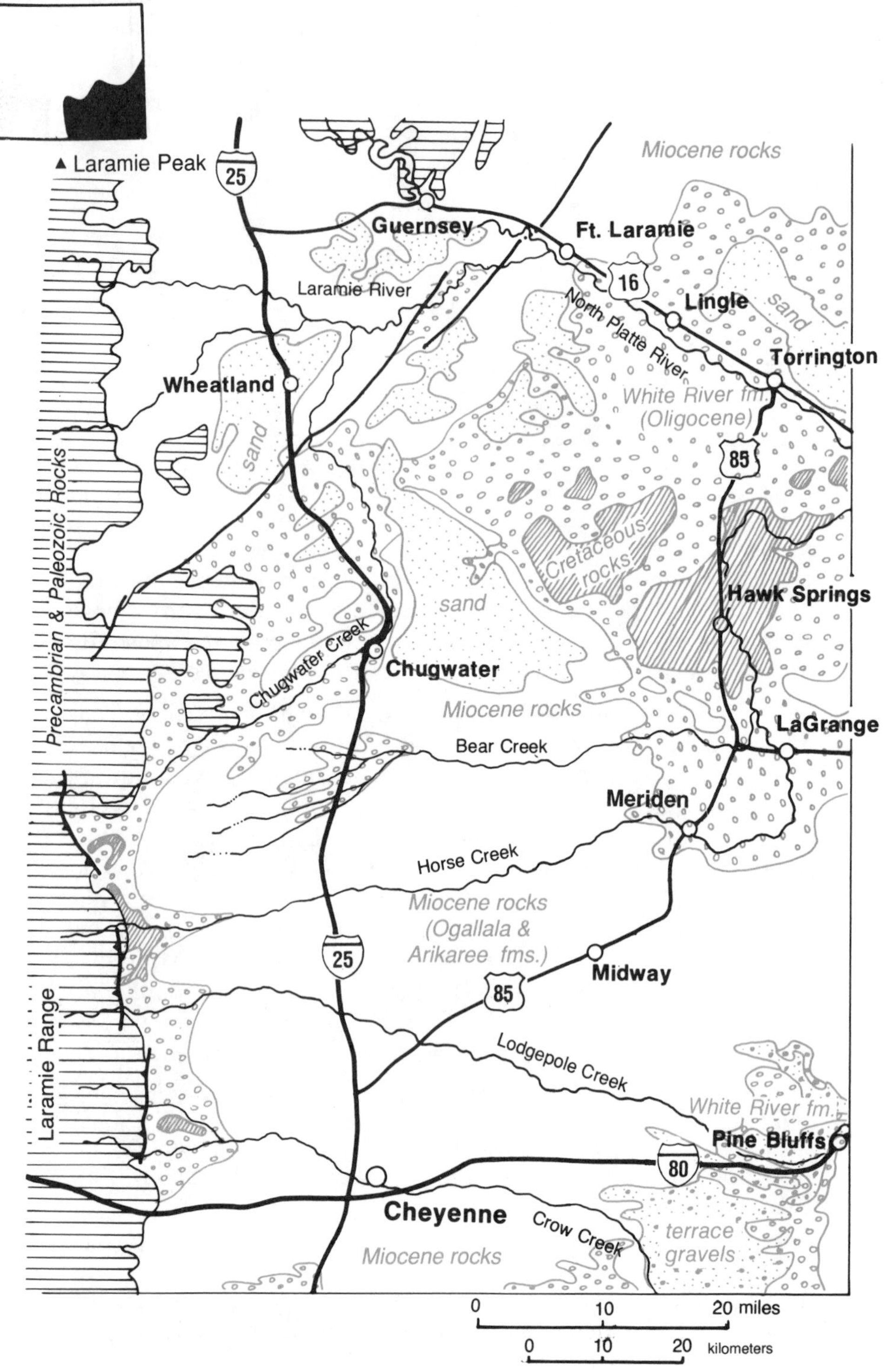

Geologic map of southeastern Wyoming.

Interstate 25: Cheyenne—Casper
184 mi./296 km.

Cheyenne is the capital of Wyoming and home of the famous Cheyenne Frontier Days Rodeo. The town began in 1867 as a construction camp for the Union Pacific Railroad, then served the Cheyenne-Deadwood Stage, hauling passengers, mail, and freight from the railroad to the gold-rich Black Hills.

Interstate 25 north from Cheyenne parallels the Laramie Range to the west. Some of the youngest sedimentary rocks in Wyoming crop out between Cheyenne and Casper. These young Cenozoic sediments were eroded from mountains to the west and deposited by rivers as a blanket of sandstone, claystone, and conglomerate over the high plains. As you drive over these high, rolling plains, you pass over ancient alluvial fans that were deposited millions of years ago. Today, rivers are not depositing sediments across Wyoming, but are instead eroding and dissecting the landscape to expose older layers of sedimentary rock. The interstate highway gently dips up and down over ridges and through gullies cut by the modern streams, now dissecting the topography.

Three formations are exposed between Cheyenne and Wheatland, from youngest to oldest, the Miocene Ogallala formation, Miocene Arikaree formation, and Oligocene White River formation. All three are composed of light-colored, tuffaceous claystone, sandstone, and lenses of conglomerate. Tuffaceous refers to volcanic ash from distant volcanos, that was carried across Wyoming by westerly winds, and deposited with river muds and sands. The Ogallala formation, named after a band of Sioux Indians, covers the high plains around Cheyenne and extends to the Nebraska and Colorado state lines on Interstate 80 and Interstate 25, respectively. The Ogallala formation gathers rainfall and snow-melt in this area, and carries it underground to the east to become a major ground water aquifer beneath the Great Plains. The underlying Arikaree and White River formations are exposed along incised stream channels and roadcuts from about 25 miles north of Cheyenne to Wheatland.

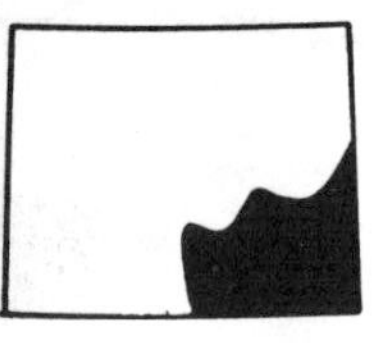

Section at Chugwater.

Iron Mountain is along the east base of the Laramie Range, about 35 miles northwest of Cheyenne; it was the site of iron mining in the past. The reddish-black iron ore came from Precambrian rocks.

At Chugwater, picturesque buttes overlooking Chugwater Creek and the town are composed of conglomerate, sandstone, siltstone, and claystone of the Oligocene White River and Miocene Arikaree formations. Chugwater was headquarters of the Swan Land and Cattle Company, one of the largest cattle outfits in the U.S. Their 113,000 herd of cattle ranged over 1 million acres from Rawlins to Ogallala, Nebraska, and from Cheyenne to the North Platte River. The company operated for 70 years; headquarters buildings in town are now a historical landmark. Among many stories for the origin of the name "Chugwater," the most popular tells of the Indian hunting practice of stampeding buffalo over the cliffs into the water below — the spot became known as "water at the place where the buffalo chug."

Interstate 25 follows Chugwater Creek for a few miles north of Chugwater, and the north-bound traveller has a first view of Laramie Peak on the distant horizon. Laramie Peak, the highest point in the range at 10,274 feet, can be seen for 100 miles from the east. It was a major landmark on the Oregon Trail.

The Laramie River Power Station can be seen east of Wheatland. This 1.65 million kilowatt, coal-fired, electric generating plant is

Laramie Peak, Laramie Range, from I-25, 15 miles north of Wheatland.

owned and operated by the Missouri Basin Power Project and serves over 2 million people in 8 states, from Montana to Iowa. The coal is shipped by rail from strip-mines near Gillette, Wyoming, an area of the state which contains incredible reserves of sub-bituminous coal. The Grayrocks dam and reservoir on the Laramie River provides water for steam and cooling of the power plant.

The Ogallala, Arikaree, and White River formations again form the majority of surface outcrops between Wheatland and Douglas, spreading a blanket of sediment over an underlying arch called the Hartville Uplift. The Miocene Ogallala formation crops out between Interstate 25 and the base of the Laramie Range to the west, whereas the underlying Arikaree and White River formations crop out adjacent to Interstate 25, around Glendo Reservoir, and south of Douglas. Just north of Wheatland, white claystones of the Arikaree formation are well-exposed along bluffs overlooking the Laramie and North Laramie rivers.

The Hartville Uplift is a structural arch that separates the Denver Basin from the Powder River Basin. The arch extends from the northeast end of the Laramie Range to the south end of the Black Hills. Precambrian, Paleozoic, and Mesozoic rocks were raised along the arch during the Laramide orogeny, but were covered by the blanket of Cenozoic sediments. Today, there is very little topography to suggest that this is a structurally uplifted region. However, south of Glendo, to the east, Permian and Pennsylvanian-aged limestones, sandstones, and red shales poke through the blanket of Cenozoic strata to form tree-covered hills along the crest of the Hartville arch. These Paleozoic strata are also exposed in a large interstate highway roadcut at the north end of the Glendo Reservoir. Glendo Resevoir was built across the North Platte River in 1958 for water storage and recreation.

The Oregon Trail followed the North Platte River from Fort Laramie west, skirting the north end of the Laramie Range. Today's route, Interstate 25, roughly parallels the trail between Glendo and Casper.

Douglas is at the southern tip of the Powder River Basin on the Paleocene Fort Union formation. The Fort Union formation consists of interbedded dark-gray shales and sandstones that dip north into the basin. White River formation crops out as interesting badlands to the east, west, and south of Douglas for several miles, forming a

Ayers Natural Bridge, Tensleep formation south of I-25 between Glenrock and Douglas.

horizontal blanket of white sandstone and claystone that covers the Fort Union and older strata. Bluffs overlooking the North Platte River at the Interstate 25 crossing, at the south end of Douglas, are composed of light-colored sandstone, siltstone, and claystone of slightly younger Miocene age.

Ayres Natural Bridge, west of Douglas, is a perfect stone bridge over LaPrele Creek that may be reached on a 5-mile paved road from Interstate 25. The bridge, formed from the Pennsylvanian Casper formation, is about 150 feet long by 50 feet high. This natural bridge formed when the meandering stream undercut both the up-stream and down-stream side of the resistant ridge. Shady elm trees provide an ideal place for picnics.

The Dave Johnston Power Plant, 6 miles east of Glenrock, is visible to the north of Interstate 25. This plant, owned and operated by Pacific Power and Light Co., was one of the first major coal-fired, steam electric plants in the western U.S. It began operation in 1958. The plant has continually been updated and currently has the capacity to produce 810,000 kilowatts of electricity. Coal to fire the boilers is mined from massive deposits of the Fort Union formation 16 miles to the northeast. These coal beds, up to 40 feet thick, lie close to the surface and are easily strip-mined. The coal is sub-bituminous grade and low in sulfur, making it less polluting.

The Dave Johnston coals lie at the southern tip of the Powder River Basin, Wyoming's greatest coal-producing region. To the north, over 12,000 square miles of land overlie immense coal beds, some up to 125 feet thick!

The Glenrock oil field has produced 2,600,000 barrels of oil from lower Cretaceous sandstones since its discovery in 1949. Pumpjacks dot the landscape on either side of the highway.

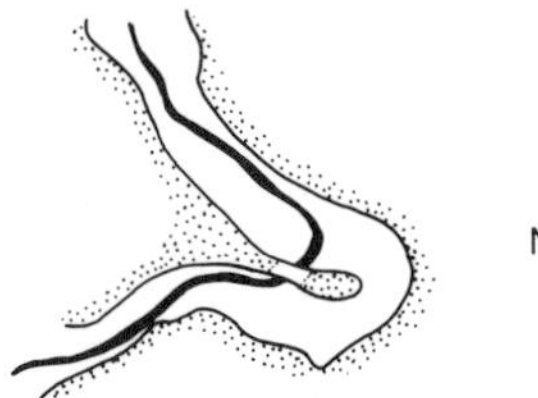

3) Undercutting by creek collapses rock, natural bridge is formed, and creek takes the short cut through the arch.

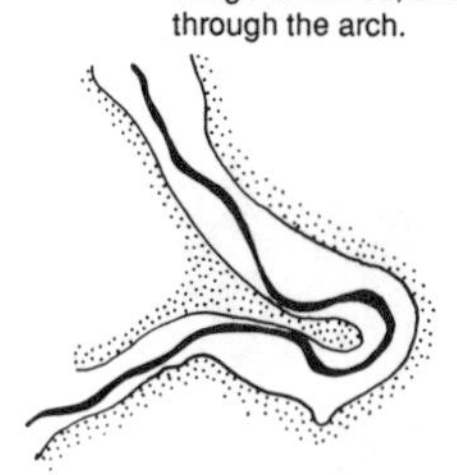

2) Creek erodes at both sides of outcrop spur.

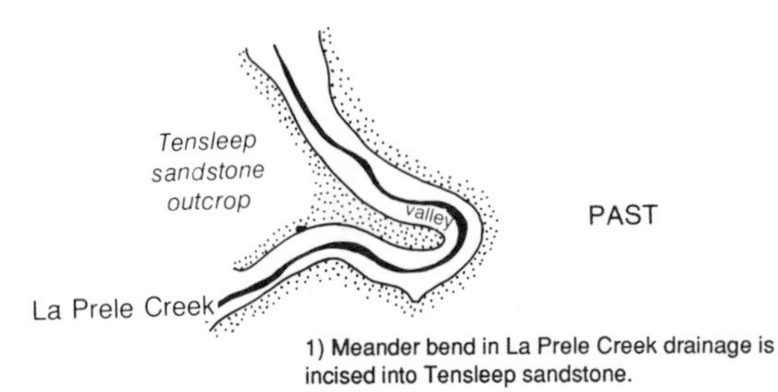

1) Meander bend in La Prele Creek drainage is incised into Tensleep sandstone.

Time sequence maps showing formation of Ayers Natural Bridge by La Prele Creek erosion.

Dave Johnston Power Plant east of Glenrock along I-25

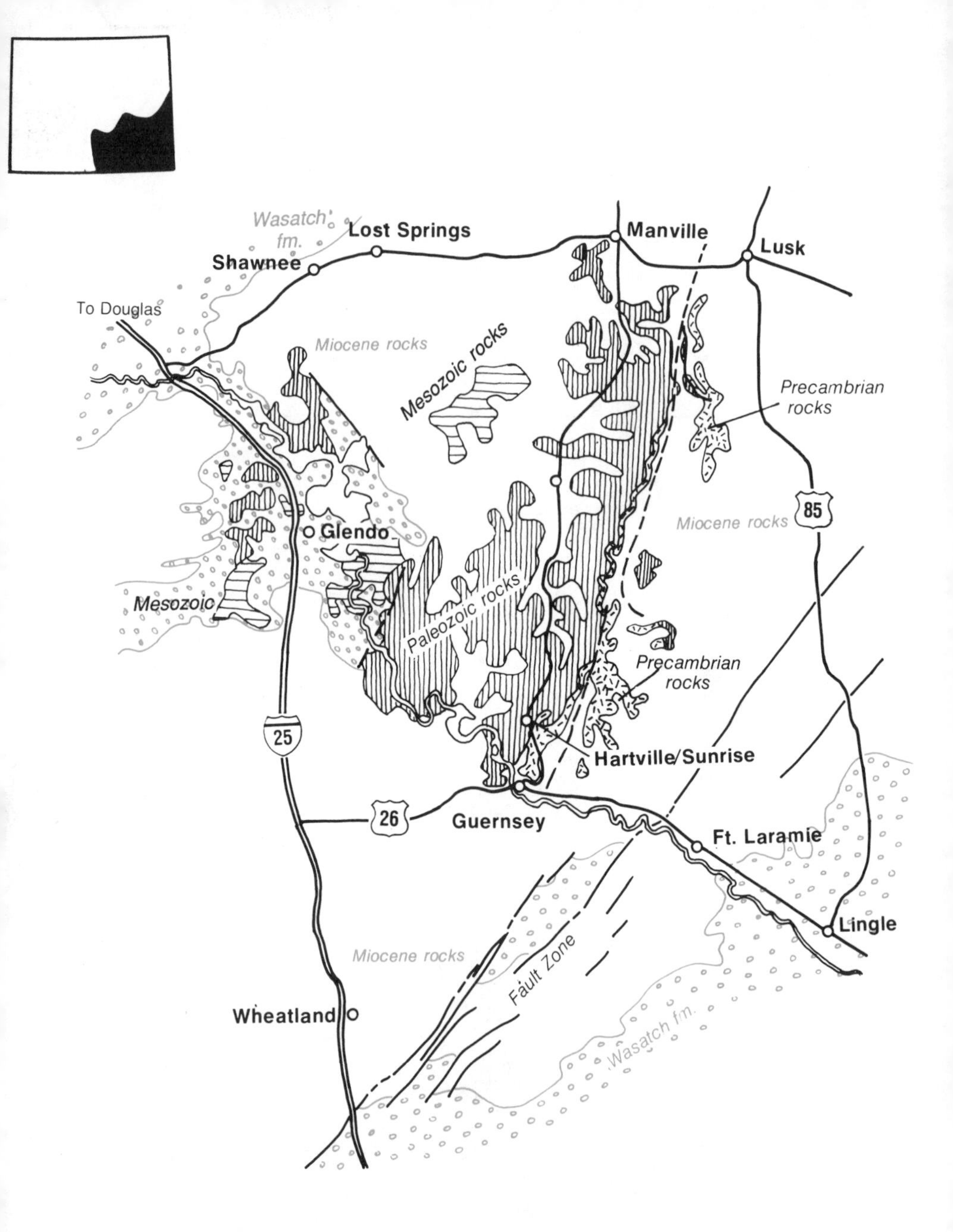

Geologic map of the Hartville Uplift area.

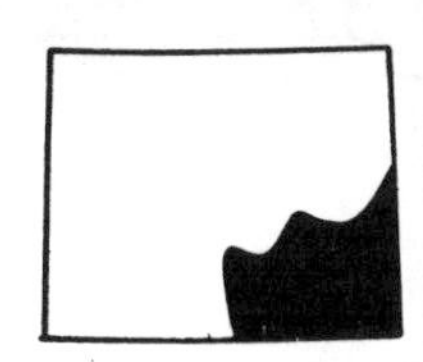

Oregon Trail wagon ruts in White River formation along US Highway 26 south of Guernsey.

US 26: Torrington—Interstate 25
48 mi./77 km.

US 26 is not a major interstate route today; it isn't even heavily travelled, but 140 years ago its route was the main artery for west-bound emigrants on the Oregon Trail. During 1850, the peak year of western migration, the Conestoga wagons were literally bumper to bumper on the trail. A pioneer's diary of 11 days during the summer of 1850 records 933 wagons passing through Fort Laramie drawn by 6,904 oxen and 544 mules, accompanied by 1,465 horses and 1,988 cows; walking alongside or riding were 393 children, 242 women, and 2,817 men! The wagons carved deep ruts into soft sandstones of the Arikaree formation near Guernsey. Register Cliff, also composed of Arikaree sandstone near Guernsey, contains the carved names of thousands of pioneers. Both spots are important historical landmarks.

Fort Laramie is the most famous fort in the Rocky Mountain west. Located at the eastern foot of the Rocky Mountains, every west-bound trapper, trader, and emigrant passed through its dusty gates. It was built at the junction of the Laramie and North Platte rivers in 1834 by fur traders William Sublette and Robert Campbell. It became increasingly important as a fur trading center after the abandonment of the trappers' rendezvous in 1840. In 1849 it was sold to the U.S.

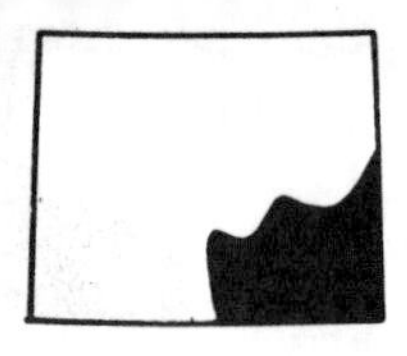

Iron Ore Mine at Sunrise, north of Guernsey, Highway 270.

government and established as a military fort for the protection of Oregon Trail emigrants. Fort Laramie was the site of the famous 1851 treaty with the Plains Indians, which gave the Indians all the land from the North Platte to Arkansas rivers, and east from the Rockies to Kansas, for "as long as the grass shall grow." The grass did not grow for long.

Geologically, the land between Torrington and Interstate 25 is relatively uncomplicated. From Torrington to two miles east of Fort Laramie, US 26 follows the North Platte River, which cuts a wide floodplain through soft, white sandstones and claystones of the White River formation. The bluffs around Fort Laramie are upper Oligocene and lower Miocene sandstones and claystones of the Arikaree formation. It blankets the Hartville Uplift to Interstate 25, except at Guernsey, where the Pennsylvanian Hartville formation pokes through. The Hartville formation contains red and white sandstones, gray dolomite and limestone, and red shale. It lies on Precambrian basement rocks north of Guernsey.

The Hartville-Sunrise Mining District is on the eastern flank of the Hartville Uplift. Steeply-dipping, folded and faulted Precambrian metamorphosed sedimentary and volcanic rocks contain hematite, an iron-oxide mineral. Indians used the red hematite in war paint long before white men entered the region. Iron mining began in 1898, using the open-pit method until the 1930s. Since then, it has been an underground mining operation. Colorado Fuel and Iron Corporation operates the Sunrise Mine. The concentrated ore is shipped to blast furnaces in Pueblo, Colorado, more than 300 miles to the south. Some copper, and minor amounts of gold and silver, have also been recovered.

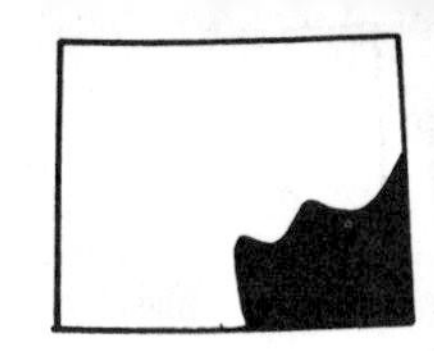

Sherman granite blocks at Vedauwoo Recreational Site, Laramie Range, I-80.

Interstate 80: Cheyenne—Laramie
49 mi./78 km.

As you travel from Cheyenne to Laramie, you leave the Great Plains behind and cross the first range of the Rocky Mountains, the Laramie Range. This crossing is made relatively easy by a geological feature called the Gangplank. The Gangplank, on the east side of the Laramie Range, is like a giant plank a few miles wide, that was laid from the plains to the top of the range, providing a low-gradient crossing over it. The plank is composed of the Miocene Ogallala formation, which gently laps onto Precambrian granite near the Harriman Road exit off Interstate 80. To the north and south, these sediments have been eroded away, but in this limited area the former level of basin-fill has remained intact. The Ogallala formation forms a flat, featureless landscape between Cheyenne and the Harriman Road. Knobs and boulders of pink granite create a rougher topography farther west.

Mining activity appears just north of the interstate, at the Harriman Road exit. Quarries in the 1.4 billion-year-old Sherman granite produce ballast, a crushed aggregate used by the Union Pacific Railroad for weighting and holding rail ties in place.

The main trunk line of the Union Pacific Railroad, like the interstate highway, follows the Gangplank between Cheyenne and

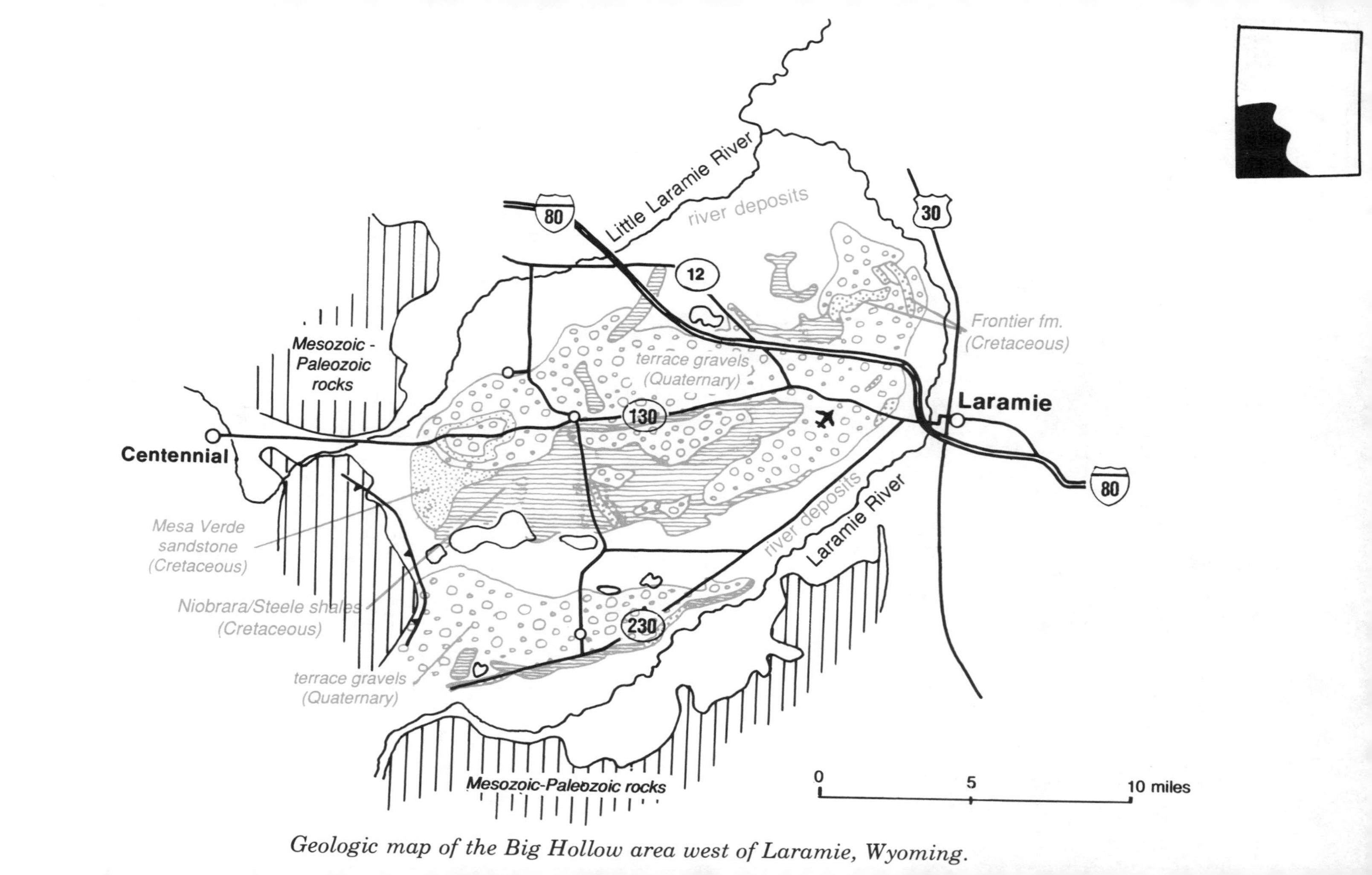

Geologic map of the Big Hollow area west of Laramie, Wyoming.

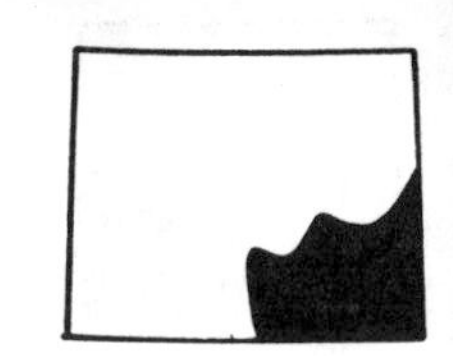

Close-up of Sherman granite (1.4 billion years old) at Vedauwoo.

Laramie. The Union Pacific, the first transcontinental railroad, was built between 1865 and 1869 at the astonishing rate of over one mile per day. It was constructed entirely with picks, shovels, and mule-drawn scoops in the hands of sweating, swearing, sun-baked men. These men built a hard reputation for raucous behavior that shocked the Victorian sensibilities of the day, as they "laid iron" across the mountains and basins of southern Wyoming. Finally, on May 10, 1869, at Promontory Point in Utah, the Union Pacific joined tracks with its western counterpart, the Central Pacific Railroad. A golden spike was driven into the last rail tie, symbolizing for many the end of the American wilderness.

The Vedauwoo Recreation Site in the Laramie Range is a worthwhile stop. Vedauwoo is Arapahoe for "earth-born," which is an appropriate name for this jumbled pile of rocks. It formed from massive outcrops of pink, Precambrian granite, 1.4 billion years old, that weathered along fractures into rounded piles of boulders, many of which are precariously balanced.

Sherman Rest Area and the Lincoln Bronze are atop 8,640 foot Sherman Hill, the crest of the Laramie Range and the highest point along Interstate 80. Until the turn of the century, the Union Pacific tracks ran over this pass, and trains always stopped to check their brakes before descending. The railroad now follows a more gradual route a few miles south.

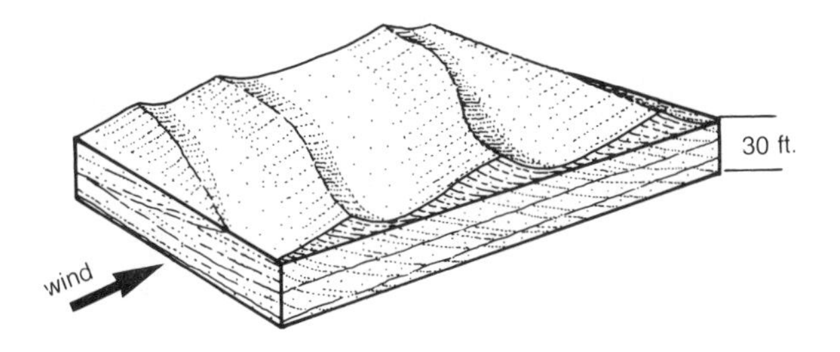

Large scale cross-bedding produced by blowing sand dunes as seen in Casper formation on Sherman Hill.

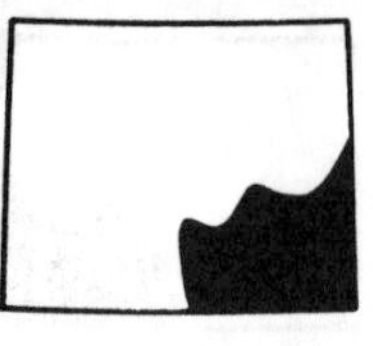

From Sherman Hill to Laramie, the highway descends through the west-dipping Casper formation, composed of interbedded, gray marine limestone and bright-red, wind-deposited sandstone of Pennsylvanian-Permian age. These rocks were deposited within, and adjacent to, a shallow marine sea that covered Wyoming about 290 million years ago, before dinosaurs roamed the land. Note the cross-bedding in the sandstones, formed as layers of sand accumulated on the steep faces of dunes.

Laramie is on the east side of the Laramie Basin, a trough between the Laramie and the Medicine Bow ranges. It was named after the Laramie River, which was named for Jacques LaRamie, a French-Canadian fur trapper. Laramie was settled in 1867 as an "end of the rail" camp during the construction of the Union Pacific Railroad. It is home to the University of Wyoming, which has an excellent geology department and museum. The museum displays a life size model of Tyrannosaurus Rex and a Brontosaurus skeleton, along with many other fossil and mineral specimens. The American Heritage Center at the university houses the Petroleum History and Research Center, an archive of many important petroleum and geological collections, including the Anaconda Minerals, Sinclair Oil, and Exxon Oil collections to name a few. The Geological Survey of Wyoming, also on campus, has many interesting publications on Wyoming geology for sale to the public.

Most of the University buildings are made of a pinkish-tan sandstone that was locally quarried from the Pennsylvanian Casper formation, which appears in roadcuts on the west flank of the Laramie Range.

Interstate 80: Laramie—Rawlins
99 mi./158 km.

If you drive this section of Interstate 80 in the summer, you'll probably see lazy herds of antelope grazing in the Laramie Basin, or be impressed by the soft, forested heights of the Medicine Bow Mountains. You may wonder about the odd looking fences next to the interstate that go nowhere. Don't be fooled by gentle summer. This highway is a battleground during the winter! Interstate 80 between Laramie and Rawlins, more than any other road in Wyoming, can be a fierce challenge to the winter traveller, and is often closed. Winds whip snow from the mountains and drive it horizontally across the landscape with the force of a hurricane. Those odd fences are baffles to slow the wind and stop the snow before it reaches the highway. It has been said that snow never melts in Wyoming; it just blows back and forth until it wears out!

West of Laramie, the floor of the Laramie Basin is Cretaceous marine shales and sandstones, overlain by Quaternary gravels that cap the higher terrain. These gravels were deposited by streams flowing from surrounding mountains as the Laramie Basin was excavated. Several small oil fields that dot the Laramie Basin are visible from the highway; they produce from the same strata that are exposed along Interstate 80 on the west side of the Laramie Range, and from lower Cretaceous sandstones. The Quealy Dome field, 19 miles west of Laramie, is on an anticlinal arch. It has produced more than 12,000,000 barrels of oil since its discovery by the California Company in 1934. This was the first oil field discovered in Wyoming with seismic techniques, which use man-made shock waves, created by either small explosions or vibrations, that travel downward into the earth. They reflect off different layers of rock and bounce back to the surface; the reflected waves are recorded on magnetic tape, then converted into a cross section of the earth below the shock. In a sense, seismology is like taking a x-ray of the earth. Petroleum geologists use seismic data to look for areas where strata arch into an anticline, which may have trapped oil or gas in layers of porous sandstone or limestone.

North of Quealy Dome exit, the Laramie Basin dropped lower than it did farther south, thus protecting part of the Eocene Wind River formation from erosion. This formation contains arkosic conglomerate, sandstone, and claystone that was deposited by rivers 50 million years ago, at the close of the Laramide orogeny. West of Interstate 80, the upper Cretaceous Mesaverde formation dips eastward into the Laramie Basin in front of the Medicine Bow Mountains. During the Laramide orogeny, Precambrian metamorphic rocks in the Medicine Bows were raised and shoved eastward along the Arlington thrust fault, which lies at the foot of the mountains to the west.

About two miles east of the Albany-Carbon County line, a stone on the hillside east of Interstate 80 marks the grave of C.S. Bengough, a

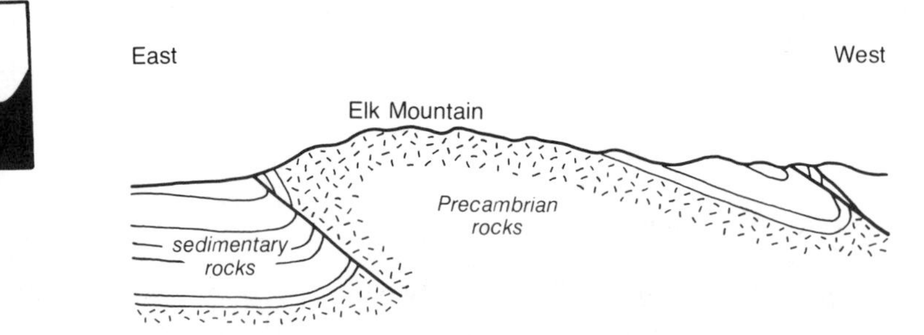

Cross section of Elk Mountain, south of I-80, west of Laramie. –Adapted from Blackstone (1980)

colorful English settler who was excommunicated to the New World by his family. Coeds from the University of Wyoming traditionally place flowers on his grave at exam time for good luck. The ridge top above the grave is capped with quartzite cobbles of the Paleocene Hanna formation that were shed from the rising Medicine Bow Mountains during the Laramide orogeny.

Between the Albany and Carbon County line and Arlington, the highway follows the north end of the Medicine Bow Mountains. The expanse of the Laramie Basin lies to the northeast. This stretch of Interstate 80 roughly follows the historic Overland Trail stagecoach and wagon route. Arlington, an important way-station on the trail, was founded in 1860 at the Rock Creek crossing. Several buildings remain, including one that housed a combination blacksmith shop, saloon, and dance hall. In 1865, Arapahoe Indians attacked a wagon train here and captured two young girls, Mary and Lizzie Fletcher, after killing their mother. Mary was eventually returned to her father, but Lizzie was never heard of again. Thirty-five years later, a white woman who was raised as an Arapahoe came to Casper with Indians from the Wind River Reservation. Mary identified this woman as her sister, but Lizzie refused to leave her Arapahoe family.

Elk Mountain anticline, about half-way between Laramie and Rawlins along I-80.

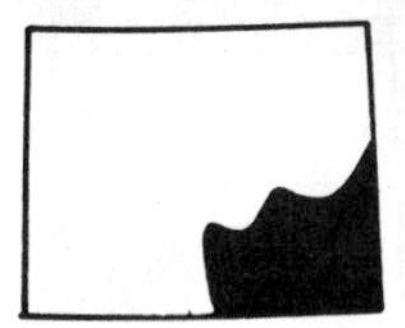

Immediately south of Arlington, the Arlington thrust fault crops out at the base of the hills; Precambrian metamorphic rocks have been uplifted by this fault to form the Medicine Bow Mountains.

West of Arlington and south of Interstate 80, the Medicine Bow River cut a broad valley between the north end of the Medicine Bow Mountains and Elk Mountain. Soft, gray, marine shales of the Cretaceous Steele formation are exposed in this valley. Sandstone beds in the overlying Cretaceous Mesaverde formation form cliffs north of the interstate highway.

Elk Mountain, just south of the interstate highway, is an isolated Laramide Uplift of Precambrian granite, faulted along its eastern and southern margins. The Mississippian Madison limestone and Pennsylvanian Casper sandstone form the gentle north flank of the mountain.

The Hanna Basin lies north of Elk Mountain. It is almost funnel-shaped, and contains more than 30,000 feet of sedimentary rocks along its north margin. The Hanna Basin is a prolific coal producer from the Paleocene Ferris and Hanna formations. These are non-marine strata deposited during the Laramide orogeny. They consist of brown and gray sandstone, shale, and conglomerate eroded from the Medicine Bow Mountains. Thick beds of coal from these formations were mined by the Union Pacific Railroad for decades to power its locomotives. The Ferris and Hanna formations are roughly correlative with the Paleocene Fort Union formation, the main coal producer in the eastern Powder River Basin.

At the north end of Elk Mountain, Interstate 80 climbs over a divide that separates the Medicine Bow and Platte rivers, called Halleck Ridge. It is formed by resistant, orange and buff-colored

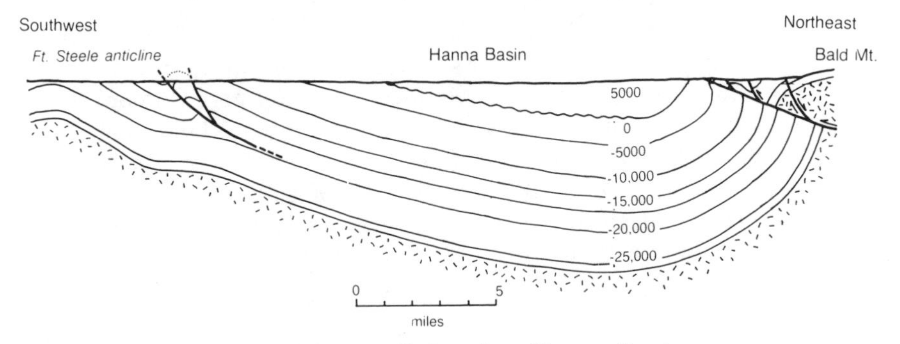

Structure of the small, but deep Hanna Basin.

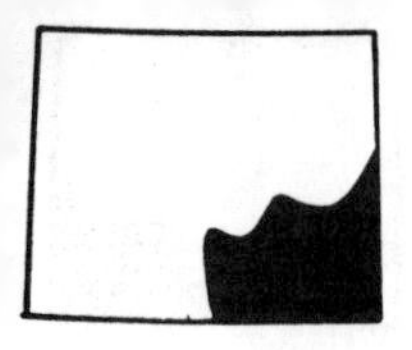

sandstone outcrops of the Mesaverde formation that have been folded into a small anticline. A sweeping view of the Saratoga Valley spreads away to the west. West of Halleck Ridge, Interstate 80 crosses the north end of the Saratoga Valley, a narrow trough separating the Medicine Bow and Sierra Madre mountains. The North Platte River flows north through the valley from its headwaters in North Park, Colorado. The valley is covered with a veneer of white, tuffaceous sandstone and claystone belonging to the Miocene Browns Park and North Park formations, which are exposed in road cuts at Walcott Junction. These formations cover the underlying, folded Mesozoic-Paleozoic strata like a blanket. They were laid down long after the Laramide orogeny.

Between Wyoming 130 and Sinclair, Interstate 80 traverses outcrops of soft, gray, Cretaceous Steele shale in the center of the Fort Steele anticline, which parallels the highway. To the north and south, sandstone beds within the Mesaverde formation dip in opposite directions on the limbs of the anticline. Like other dark Cretaceous shales in Wyoming, the Steele shale was deposited 80 million years ago in the quiet waters of a shallow seaway that extended through the western interior of North America. Dinosaurs reached their evolutionary zenith at this time, never dreaming that their mass extinction was just around the corner. Ash from volcanos to the west drifted east and settled in the Cretaceous seaway to form layers of bentonite clay. Bentonite makes the Steele shale extremely slippery and sticky when wet, leading to the nickname "gumbo-shale." The Steele formation and Fort Steele anticline were named after Fort Steele, an active Army post from 1868 to 1886 that provided protection for Union Pacific Railroad workers.

Low ridges south of Interstate 80 at Sinclair are sandstone beds within the Cretaceous Frontier formation that encircle the Grenville dome, a classic anticline that for some reason contains no oil. Sinclair was initially named Parco, for Producers and Refiners Corporation, the original operators of the refinery. The refinery was built along the main line of the Union Pacific Railroad in the 1920s by Parco, and was acquired by Sinclair Oil Company in 1934. It has been modernized over the years and today is the biggest refinery in Wyoming, and one of the biggest in the Rockies.

Black shales and sandstones of the Cretaceous Frontier formation are exposed in road cuts just west of Sinclair. These beds dip east off the east flank of the Rawlins Uplift. The Frontier sandstone is an important oil reservoir rock throughout Wyoming, and the interbedded black shales testify to the carbon-rich, oil-producing quality of the formation.

The Rawlins Uplift defines the eastern margin of the Great Divide Basin, the western end of southeastern Wyoming, and the beginning of the desert basins to the west. It is a small Laramide Uplift that has many of the characteristics of larger Wyoming ranges. Precambrian basement rocks are exposed in the core of the uplift, and its flanks are composed of outwardly-dipping Paleozoic and Mesozoic strata. A thrust fault dips beneath the uplift along the west and south sides. Rawlins lies at its south end.

US 30/287 and Wyoming 487 Laramie—Casper

150 mi./240 km.

US 30/287 follows the meandering Laramie River north for about 20 miles across the Laramie Basin. The river eventually flows through a narrow canyon it cut across the central Laramie Range. By the end of Tertiary time, the Laramie Basin had filled with sediments which lapped over the adjacent Laramie Range. The Laramie River established its course over the range on top of these sediments, then cut a deep canyon during the present cycle of basin excavation that began around 2 million years ago. This process of canyon-cutting is called "superposition," and it accounts for many of the stream-cut canyons throughout Wyoming.

Red shales immediately north of Laramie belong to the Triassic Chugwater formation, a non-marine unit deposited after retreat of the Paleozoic seaways, before the Jurassic Sundance seaway flooded the region. Farther north, Quaternary river gravels cover the basin floor to the town of Bosler.

Gray shales of the upper Cretaceous Steele formation crop out at Bosler. The Eocene Wind River formation unconformably covers these shales one mile north of Bosler, and forms the floor of the Laramie Basin for the next eight miles. The Wind River contains claystones, sandstones, and lenses of river-deposited conglomerate. About 15 miles north of Bosler, the railroad and highway cut through

a prominent ridge formed by the Pine Ridge sandstone member in the upper Mesaverde formation. The Pine Ridge sandstone was deposited as an offshore sand bar within the Cretaceous seaway that covered the western interior of North America. The underlying Steele shale is exposed along the Rock Creek Valley north of the Pine Ridge sandstone.

Wyoming 30/287 traverses the Steele shale for ten miles north of the town of Rock River. The Laramie Range and Laramie Peak, at an elevation of 10,274 feet, frame the northeast horizon. Located north of Rock River, the old Union Pacific Railroad Station of Wilcox was the site of Butch Cassidy and his Hole-in-the-Wall gang's first renowned train robbery. On a June morning in 1899, six masked men flagged down a train, dynamited the baggage and express car, and escaped with thousands in bills and bank notes. The ensuing chase was popularized by Hollywood and is now legendary.

From ten miles north of Rock River to Medicine Bow, the highway cuts through whitish chalks of the Niobrara formation, then parallels south-dipping sandstone beds of the Frontier formation on the south flank of Como Bluffs anticline. Like the Pine Ridge sandstone, these sands, shales and chalks were also deposited in the Cretaceous marine seaway. Niobrara chalks are roughly equivalent in age to the chalk cliffs of eastern England. Chalk is a sedimentary rock composed of microscopic marine organisms with calcite shells.

Don't miss the dinosaur bone cabin just north of the highway! These bones were dug from the Jurassic Morrison formation on the north side of Como Bluffs anticline. Como Bluffs in a legendary dinosaur fossil site and was the stage for a notorious paleontological feud in the 1870s. The Como Bluffs story is told in the chapter introduction.

About one mile west of the bone cabin, US 30/287 crosses silver-gray shales of the Cretaceous Mowry formation exposed on the plunging west nose of Como Bluffs anticline, then crosses the floodplain of the Medicine Bow River. The town of Medicine Bow was immortalized by Owen Wister's novel, *The Virginian*, published in 1902. The book, based on Wister's experiences in Wyoming, is generally credited as being the first western novel, and established a new literary genre. The historic Virginian Hotel is a landmark in Medicine Bow.

US 30/287 continues west of Medicine Bow across the Hanna Basin to join Interstate 80. The Hanna Basin is one of Wyoming's major

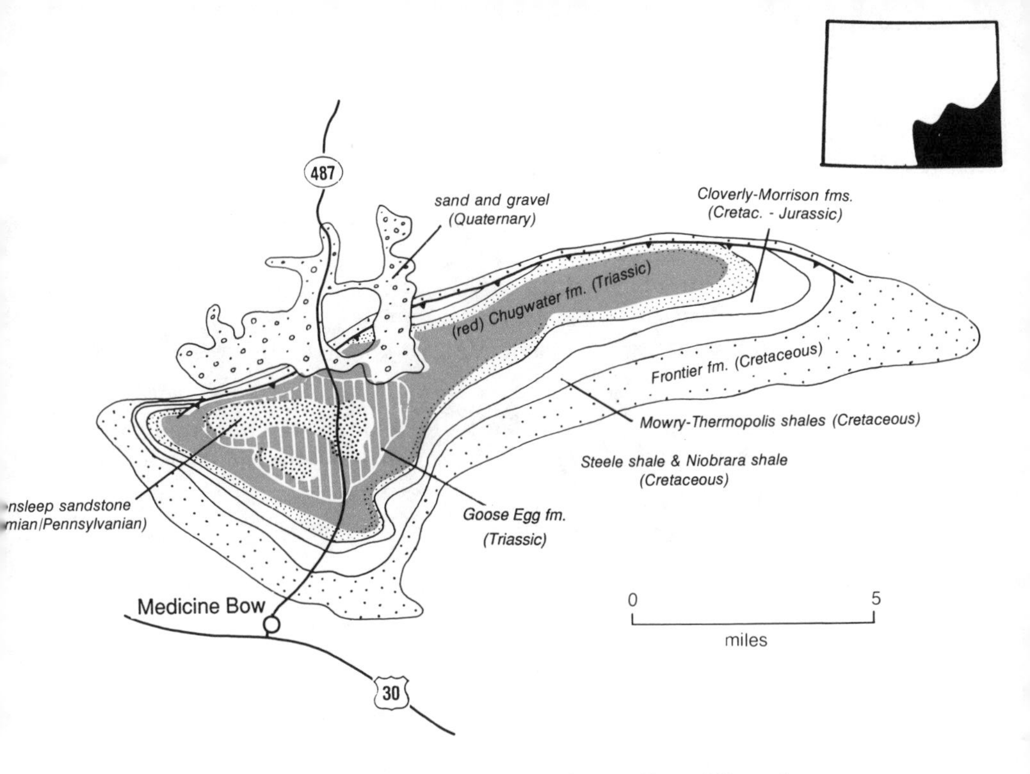

The Flat Top Anticline north of Medicine Bow, Wyoming.

coal-producing basins.

Wyoming 487 to Casper crosses Flat Top anticline immediately north of Medicine Bow. Like Como Bluffs anticline, Flat Top has a gently dipping south flank and a very steep north flank. It moved northward on a thrust fault exposed along its north flank. Resistant sandstones of the Pennsylvanian Casper formation form a prominent hill in the center of Flat Top, 5 to 6 miles north of Medicine Bow.

North of Flat Top anticline, Wyoming 487 skirts the east end of the Freezeout Mountains. The Freezeout Mountains contain uplifted Mesozoic and Paleozoic strata on the east end of the Shirley Mountains, which in turn lie at the eastern end of the larger Granite Mountains Uplift. The name, "Freezeout," may have come either from a mail carrier who was caught by a storm and wrote "froze out" on a rock, or from a group of English hunters who were similarly caught in a storm in the 1870s. The highway follows east-dipping outcrops of the Cretaceous Frontier formation around the east and northeast margins of the uplift.

Wyoming 487 forks with Wyoming 77 around the Shirley Basin. Both routes rejoin at the north end of the basin. The Shirley Basin is a

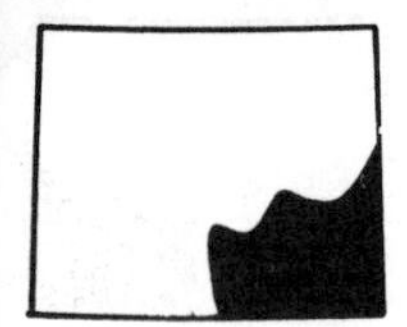

Cliffs of Oligocene White River formation at the north end of the Shirley Basin along highway 487 south of Casper.

small depression between the Laramie Range and Freezeout or Shirley mountains. Its floor is full of flat-lying sandstones of the Eocene Wind River formation, which contain large deposits of uranium. Like other uranium deposits in Wyoming, the uranium leached from Precambrian granitic rocks and moved in percolating ground waters into porous sandstones of the Wind River formation. The Shirley Basin contains large reserves of uranium, which are mined from open pits. The north end of the Shirley Basin is covered by white, tuffaceous claystones and sandstones of the Oligocene White River formation, dated at 31 to 35 million years old. The White River formation forms impressive cliffs to the north as Wyoming 487 descends to join Wyoming 77 at the north end of the basin.

North of the junction with Wyoming 77, Wyoming 487 follows the Cretaceous Steele shale for 14 miles along the Stinking Creek Valley. High, white mesas east and west of the highway are capped by the Oligocene White River and Eocene Wind River formations, which blanket the underlying Cretaceous strata. These rocks demonstrate a striking contrast in the geologic history of Wyoming, from the Cretaceous marine, black shales to the overlying, non-marine, white Tertiary strata.

As Wyoming 487 turns west to join Wyoming 220, it cuts downsection through the Niobrara, Frontier, Mowry-Thermopolis, and Cloverly formations that dip south off Muddy Mountain. The Mowry formation is a distinctive, hard, silver-gray shale that contains abundant fish scales; the Thermopolis contains black, soft, fissile shales that easily split into thin layers along bedding planes. The Cloverly

formation contains variegated claystones. It is exposed directly east of Wyoming 487. Bright red Triassic shales of the Chugwater formation appear on the north face of Muddy Mountain, farther east.

Wyoming 487 joins Wyoming 220 between Alcova Reservoir and Casper. Wyoming 220 follows the Platte River through "Bessemer Bend" around the west side of Casper Mountain. This short section is covered in the central Wyoming chapter.

THE SNOWY RANGE ROAD

Wyoming 130: Laramie—The Saratoga Valley
71 mi./114 km.

The Snowy Range road over the Medicine Bow Mountains is one of the most picturesque highways in Wyoming. It is open for travel during the summer and fall, but is closed by snow the rest of the year. The crest of the Medicine Bow Mountains is called the Snowy Range because of the sugary-white, Precambrian quartzites that reflect sunlight and look like snow, even in late summer.

Wyoming 130 heads west from Laramie across the Laramie Basin. Cretaceous marine black shales are overlain by gravels. These gravels were deposited during Pleistocene time by streams that carried quartzite from the Medicine Bow Mountains. West of the airport, the Big Hollow Basin lies immediately south of the highway. Because

Glacial cirque, Tarn Lake, Talus Slopes and moraines at the crest of Medicine Bow Mountains between Centennial and Saratoga on Wyoming 130.

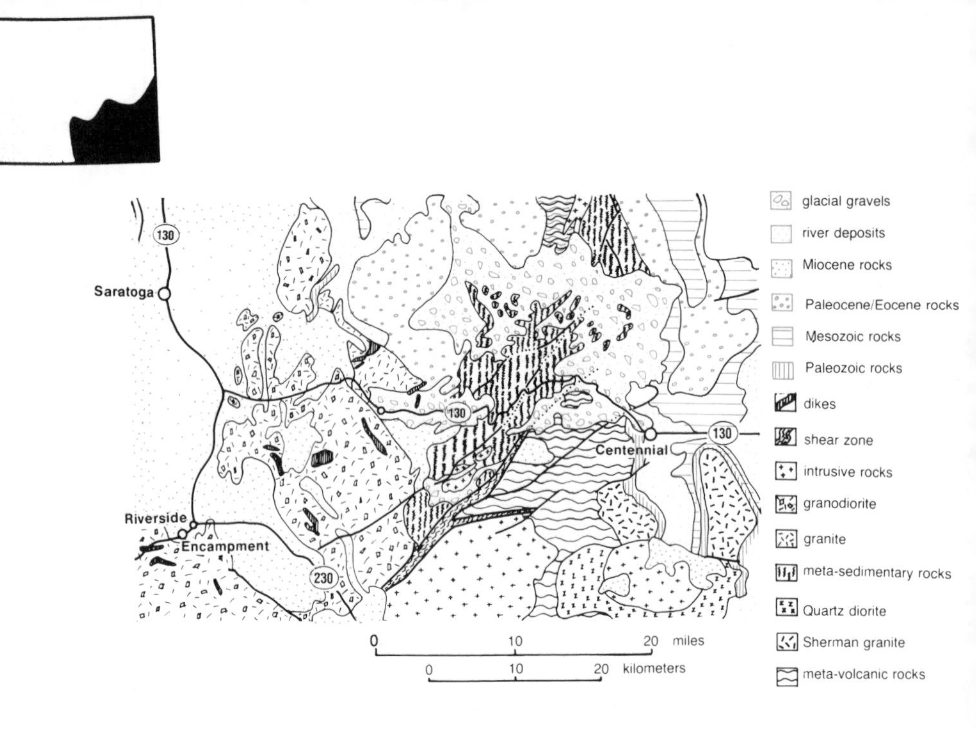

Geologic map of Snowy Range between Saratoga and Centennial.

no streams run through the Big Hollow, geologists believe it formed by wind deflation during the ice age (see page 42). Glaciers flowed from the mountains onto the basin floor at Centennial, due west from Big Hollow, and winds coming off the glacial ice scoured out the Big Hollow.

Sheep Mountain is the prominent timbered ridge south of the highway on the west side of the basin. Sheep Mountain is a mass of Sherman granite, 1.4 billion years old. A thrust fault along the eastern base of the mountain shoved the ridge upward and eastward during the Laramide orogeny.

The Centennial Valley lies between Sheep Mountain and the main front of the Medicine Bow Range. It is a nearly symmetrical syncline — the Paleozoic-Mesozoic strata are inclined into the valley at nearly the same angle on both sides. To the northwest, Corner Mountain forms a prominent corner of the Medicine Bow Mountains front where Precambrian basement rocks moved over Mesozoic strata along the Arlington thrust fault.

From Centennial, Wyoming 130 climbs into the Medicine Bow Mountains through dark outcrops of biotite gneiss that also contains large crystals of feldspar and quartz. Just east of the Brooklyn Lakes road, Wyoming 130 crosses the Mullen Creek-Nash Fork shear zone into 1.7 billion year old metamorphosed sedimentary rocks. These rocks, the Libbey Creek group of formations, were turned on-end so they now overlie much older basement rocks northwest of the shear zone. The Libbey Creek group contains schist, quartzite, quartz-pebble conglomerate, meta-dolomite, and marble. Some of the dolomite beds contain large, round balls of fossilized algae, some of the earliest life on this planet. The most distinctive formation within the group is the white Medicine Peak quartzite, which forms the range crest. Medicine Bow Peak caps the range at 12,013 feet. Numerous alpine lakes fill glacially-carved depressions along the eastern base of the crest.

Recent glacial sediments overlie 2.5 to 3.0 billion year old gneisses and granite along Wyoming 130 on the upper west side of the Medicine Bow Mountains. From the base of the range to Saratoga, the late Miocene North Park formation covers the floor of the Saratoga Valley. It consists of white to greenish-gray sandstone and claystone. The Saratoga Valley is a narrow trough that separates the Medicine Bow and Sierra Madre Mountains. The North Platte River flows northward through the valley from North Park, Colorado.

DIAMONDS ON THE STATE LINE

US 287:
Fort Collins, Colorado—Laramie
66 mi./106 km.

This section of highway into Wyoming is highlighted because of its heavy travel and the unique occurrence of diamonds along the Wyoming/Colorado state line. Between Fort Collins and Ted's Place, US 287 follows the Cache le Poudre River as it cuts through east-dipping, upper Paleozoic and Mesozoic strata. North of Ted's Place, the highway follows the general strike of these strata to Livermore, north of which it crosses a small graben containing Permo-Pennsylvanian sandstones and limestones. The road encounters Precambrian granitic rocks at Virginia Dale and crosses them all the way into Wyoming. These granites are 1.4 billion years old and correlate with the Sherman granite farther north in the Laramie Range.

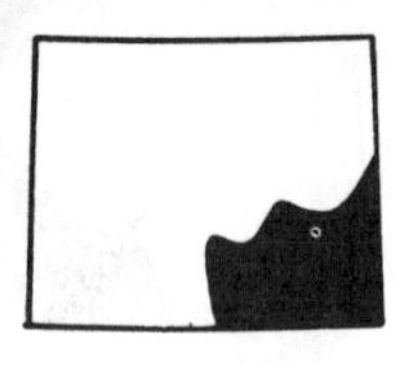

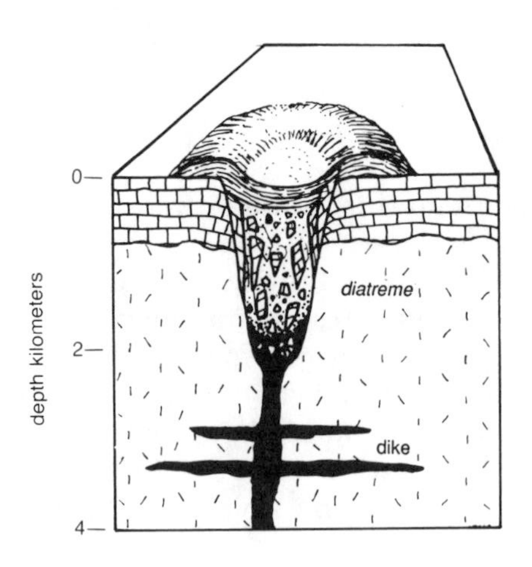

Structure of a diamond-bearing diatreme. Explosive magma rips upward through crustal rocks along fracture systems, producing a cylindrical pipe filled with brecciated rock.
–Adapted from McCallum and Mabarak (1976,

Diamonds were discovered along the Wyoming-Colorado state line in 1975. They are in structures called kimberlite diatremes. Kimberlite comes from the Earth's upper mantle, bringing diamonds with it. Diatremes are cylindrical pipes that extend to great depths in the crust and are filled with thoroughly broken or brecciated rock, mostly kimberlite. Kimberlite diatremes are believed to be deep-seated volcanic pipes through which a fluidized gas-solid mixture was injected from the Earth's mantle under high pressure. They are not typical volcanic necks, but rather formed as ascending gases brecciated the surrounding rock. Many diatremes seem to merge at depth into dikes, suggesting that hot gas-rock mixtures welled up into deep crustal fractures and, when the overlying pressure decreased enough, the gas-rock mixture rapidly expanded and blew to the surface through diatreme pipes to form small craters. Small diamonds were carried to the surface with the kimberlite fragments.

South Africa contains the most famous diamond-bearing kimberlite diatremes. The Soviet Union is also a major diamond exporter. Kimberlite rocks have been found in many parts of North America, but only two areas are now known to contain diamonds: that along the Wyoming-Colorado state line and a district in Pike County, Arkansas. The Wyoming-Colorado diatreme district covers about 50 square miles. The diatremes trend roughly north-south, parallel to the Colorado Front Range. They contain fragmented blocks of Ordovician and Silurian limestone and dolomite that were broken during forceful intrusion of the diatreme.

Most of the diamonds found thus far are very small, less than 1 millimeter across, although larger stones may exist. Diamond mining has the largest waste rock to product ratios of any commodity. For

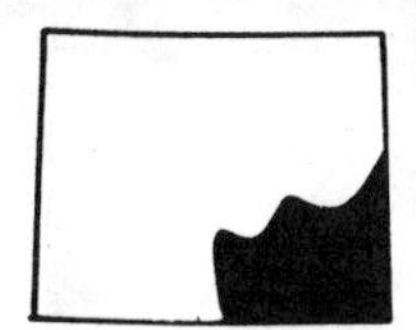

example, the diamond-rich kimberlite diatremes of South Africa yield about one carat per ton of rock (2 grams/ton), or one part diamond per 4.5 million parts of waste rock. Obviously, a tremendous amount of ore must be processed before that stone can be worn on someone's left hand! Also, sizeable gem-quality stones must be present to make diamond mining economic. The potential profitability of a mining operation in this region is doubtful because diamonds are not abundant and the cost of mining them is high.

North of the state line, US 287 drops into the Laramie Basin on the west side of the Laramie Range. Sandstone and limestone beds of the Pennsylvanian Casper formation lie directly on Sherman Granite at Tie Siding, established in 1868 to supply railroad ties to the Union Pacific. Farther north at Red Buttes, red shales of the Permian Satanka and Triassic Chugwater formations crop out.

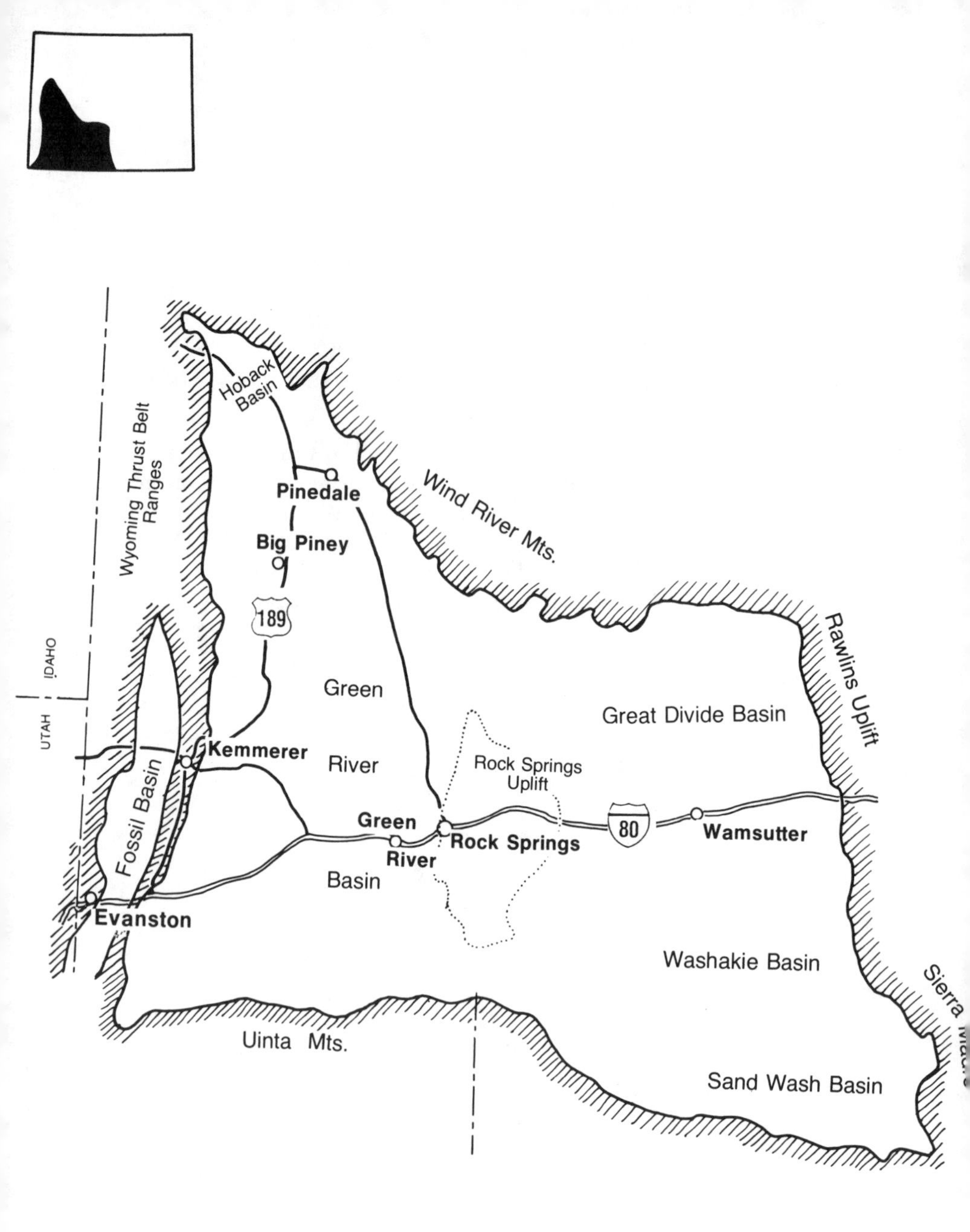

Basins, ranges and towns of southwest Wyoming.

II
Desert Basins of Southwest Wyoming

INTRODUCTION

The Greater Green River Basin is a vast desert region that occupies an enormous part of southwestern Wyoming. As you drive across southwestern Wyoming you may think there is no end to the stark, wind-blown landscape. It is a region that most people rush through, rather than pause and enjoy. However, the real beauty of the Green River Basin lies in its very vastness, its emptiness. Few places are left in this country that equal this unspoiled openness. At one spot between Rock Springs and Pinedale, you can stand on a bluff and see both the Wind River Mountains and Uinta Mountains with no evidence of man's presence in between. The famous geologist and explorer of the American West, John Wesley Powell, best described the beauty and solitude of this region in 1869:

> Standing on a high point, I can look off in every direction over a vast landscape, with salient rocks and cliffs glittering in the evening sun. Dark shadows are settling in the valleys and gulches, and the heights are made higher and the depths deeper by the glamour and witchery of light and shade. Away to the south the Uinta Mountains stretch in a long line, — high peaks thrust up into the sky, and snow fields glittering like lakes of molten silver, and pine forests in somber green, and rosy clouds playing around the borders of huge, black masses; and heights and clouds and mountains and snow fields and forests and rocklands are blended into one grand view. Now the sun goes down, and I return to camp.

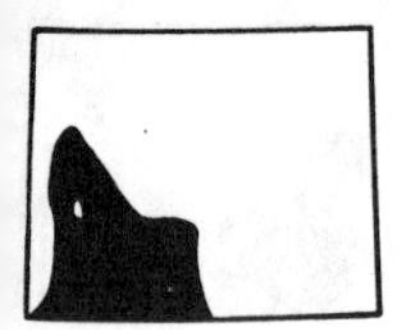

The vastness of the Greater Green River Basin is evident when placed in a geographic perspective. The area extends from the overthrust belt on the west to the Rawlins Uplift on the east, and from the Uinta Mountains to the Wind River and Gros Ventre Mountains. It is a composite basin made up of smaller basins and uplifts. The Rock Springs Uplift lies in the central part of the basin and separates it into eastern and western halves. The eastern half includes the Washakie Basin and the Great Divide Basin (also called the Red Desert Basin); the western half includes the Bridger Basin, Green River Basin proper, and the Hoback Basin.

The Greater Green River Basin is named after the Green River, which originates in the snow fields of the northern Wind River Mountains. The Crow Indians called the river "Seeds-ka-dee," meaning "Sage Hen River," and fur trappers later christened it Rio Verde, meaning "River Green." It was formally named the Green River in 1823 by Gen. William Ashley either because of its color and the greenish-tan hue of rocks along its banks, or perhaps after a fur trading associate named Green. The Green River flows south into Utah, eventually joining the Colorado River.

The Greater Green River Basin was the heart of the fur trapping country in the era of the Mountain Men. The "Seeds-ka-dee" and its tributaries were the greatest beaver waters ever known and many summer rendezvous of trappers, traders, and Indians were held in this area. It was also a thoroughfare for west-bound emigrants of the mid-1800s. Today it holds a wealth of natural resources that include oil, natural gas, coal, oil shale, and trona.

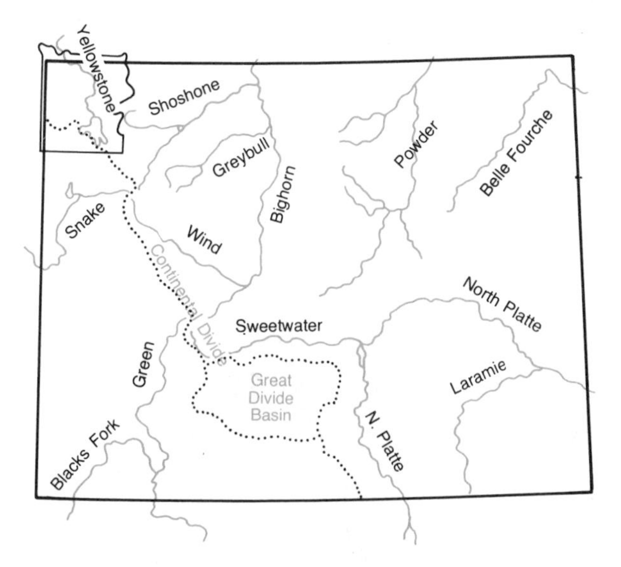

Interstate 80: Rawlins—Rock Springs
108 mi./173 km.

Interstate 80 parallels the Wamsutter arch between Rawlins and Rock Springs. The arch separates the southern Great Divide Basin from the northern Washakie Basin. The Great Divide Basin covers approximately 3,500 square miles and ranges in elevation from 6,500 feet near the center to more than 8,000 feet along the margins. The center of the basin is very arid and windy and contains sand dune deposits, the Killpecker Dune Field and saline playa lake beds. The Great Divide Basin is so named because it is topographically closed and almost no precipitation can leave as surface runoff. If this was an area of high precipitation, a large lake would probably form in the center of the basin! The Continental Divide splits when it encounters the northwest corner of the basin, and wraps around both the east and west margins of the basin; therefore, you cross the Continental Divide twice between Rawlins and Rock Springs. The Great Divide Basin is also sometimes called the Red Desert Basin for the red soil derived from Eocene formations that cover the basin's floor. This is one of the most important pronghorn antelope ranges in the state.

The Washakie Basin covers about 2,600 square miles enclosed within high rock rims on the north and southwest margins. Elevations range from 6,000 feet to 8,700 feet at Pine Butte on the west margin. The basin drains west into the Green River and south into the Little Snake River, which flows into the Green.

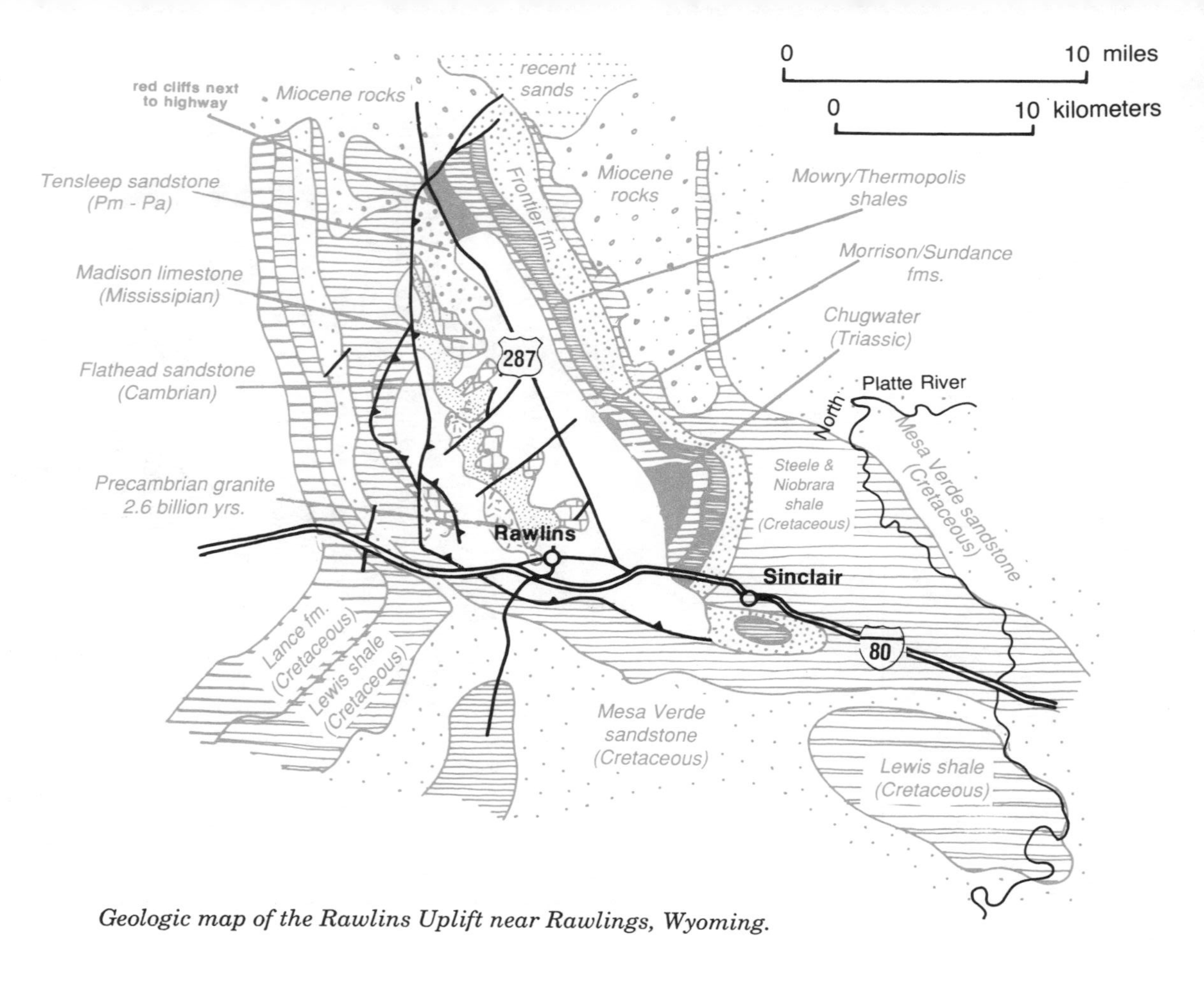

Geologic map of the Rawlins Uplift near Rawlings, Wyoming.

Rawlins lies south of the Sweetwater jade field and just east of the Great Divide (or Red Desert) Basin, famous for its gemstones. Numerous rock shops and jewelry stores in the city feature articles made of Wyoming jade, agate, and petrified wood.

Rawlins is on the south end of the Rawlins Uplift, a large thrust-faulted anticline that formed during the Laramide orogeny. Precambrian granite and Paleozoic strata in the core of the anticline compose the hills immediately north of Rawlins. The road crosses the Rawlins thrust fault about two miles west of town, at the base of high hills where Paleozoic and Mesozoic strata have been shoved to the west over younger Cretaceous rocks. For the next several miles to the west, Interstate 80 traverses Cretaceous strata that dip about 20 degrees westward into the Great Divide Basin. The Great Divide Basin is basically a large syncline between the Rawlins Uplift and the Rock Springs Uplift; therefore, strata near Rawlins dip westward into the basin, whereas those near Rock Springs dip eastward.

Interstate 80 gently curves to the southwest approximately 7 miles west of Rawlins. From this point to Riner Road, Interstate 80 crosses the Paleocene Fort Union formation, composed of brown to gray sandstone, gray shale, and thin beds of coal. From Riner Road to Creston Junction, the highway traverses an upland surface on the Eocene Battle Springs formation, which contains sandstone, mudstone, and conglomerate. Both units were deposited by streams during the Laramide orogeny when the basin was slowly subsiding between adjacent, rising mountains. The Battle Springs formation merges to the south with finer-grained sediments of the Wasatch formation that were deposited farther from the mountains. Paleocene strata are mostly drab-colored, brown or greenish-gray, whereas Eocene strata are lighter with interbedded red layers — variegated colors.

The hills rising in the distance to the south are composed of Cretaceous strata that dip northwest off the north end of the Sierra Madre Mountains.

Between Creston Junction and Wamsutter, Interstate 80 traverses the main body of the Eocene Wasatch formation. Sandstones, siltstones and variegated mudstones crop out as low hills. Variegated mudstones are the key to identifying the Wasatch formation throughout the basin.

Soft sediment folds in upper Cretaceous strata on the east flank of the Rock Springs uplift along I-80. These folds probably formed when the sediment was still soft and unlithified.

The town of Wamsutter was named in 1885 for a German bridge builder who was employed by the Union Pacific Railroad. "Turritella agate," named for a small Eocene snail with an elongated, tightly twisted shell, is scattered over the ground south of Wamsutter. The calcium carbonate shells were replaced with silica after deposition (a process called silicification), so they are very hard and excellently preserved.

The Wamsutter gas field, between Wamsutter and Red Desert north of the interstate, produces natural gas from sandstones in the upper Cretaceous Lewis and Mesaverde formations. Numerous gas fields extend south of Wamsutter along the eastern margin of the Washakie Basin.

Tipton, just west of Red Desert, was the site of one of the west's most famous train robberies. On August 29, 1900, Butch Cassidy and his "Hole-in-the-Wall" gang held up the Union Pacific Railroad near here. They got away with only $50.40, but did $3,000 damage to the train. This was their second U.P. robbery — their first was a year earlier at Wilcox, and the same express agent was the victim in both hold-ups! Although the law was in hot pursuit, Butch Cassidy and the "Sundance Kid" managed to evade them for some time to come.

Numerous oil and gas fields are along Interstate 80 between Red Desert and Point of Rocks; Patrick Draw, one of the largest, produces

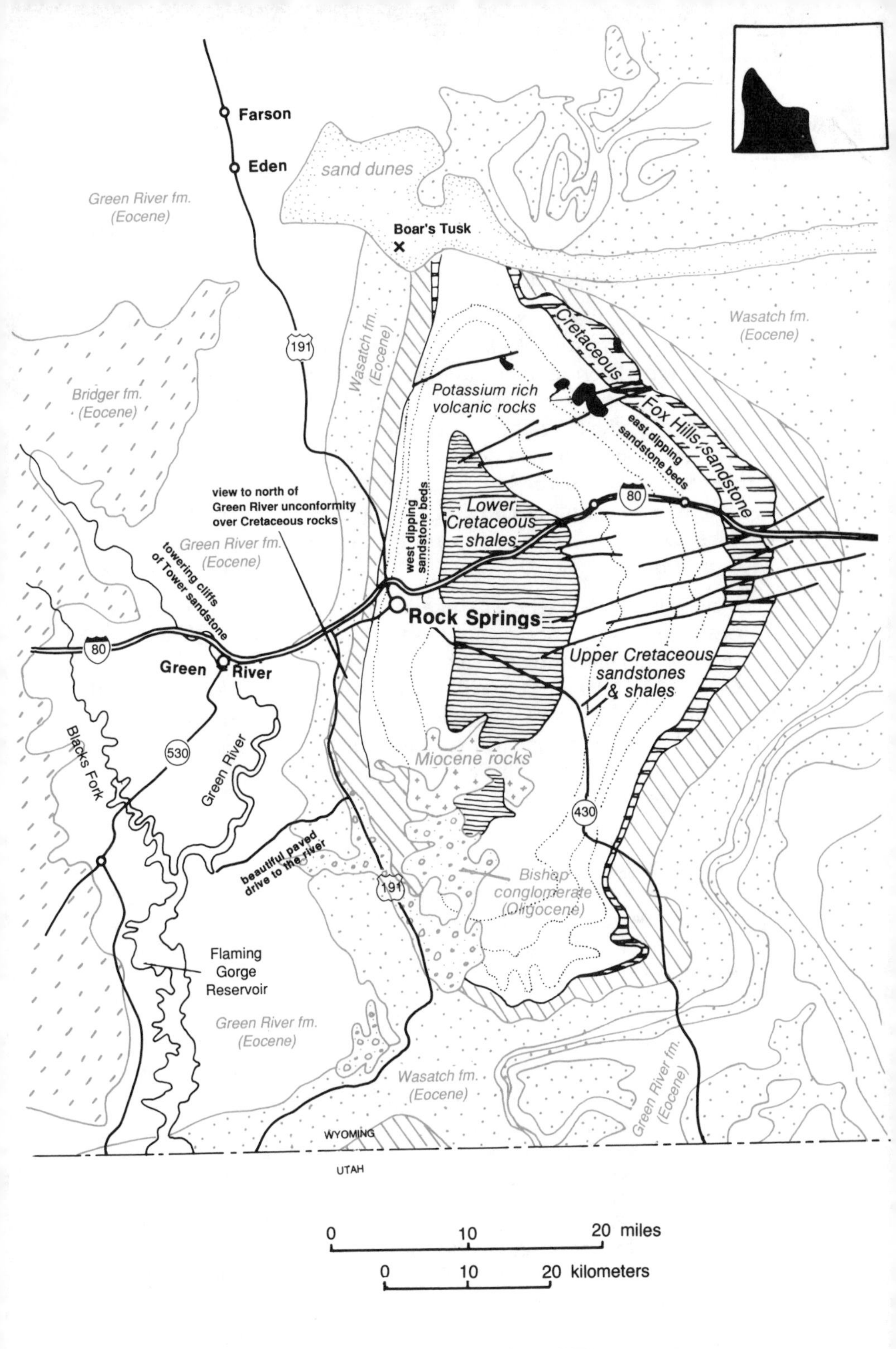

Geologic map of Rock Springs uplift and Flaming Gorge area.

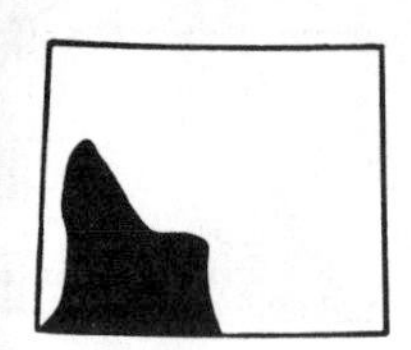

oil and gas from the upper Cretaceous Fox Hills, Lewis, and Mesaverde formations. Surface rocks in this area belong to the main body of the Eocene Wasatch formation which contains variegated sandstones and mudstones.

Interstate 80 traverses the east flank of the Rock Springs uplift in the vicinity of Point of Rocks. The east flank of the uplift is marine Cretaceous and non-marine Paleocene sandstones and shales that dip eastward 5 to 10 degrees, into the Great Divide Basin. Some of these formations produce oil and gas in the fields to the east. The white, brown, and orange sandstone beds in the Mesaverde formation are a producing horizon at the Patrick Draw field. These sandstones were deposited in a shallow seaway that covered most of Wyoming during Cretaceous time. Massive cliffs are formed by these sandstones just west of Point of Rocks.

Point of Rocks served as a stage station along the Overland Trail during Pony Express times, and was the main railroad stop in the late 1800s for miners heading north to the South Pass gold fields. North of here is the Jim Bridger Power Plant, built during the 1970s by Pacific Power and Idaho Power Companies. Low-sulfur, sub-bituminous coal from the Fort Union formation mined nearby is burned to heat boilers, creating steam that drives the turbines to generate electricity. The plant creates 2,000,000 kilowatts of electricity for customers in Wyoming and Idaho.

At the Superior exit, roadcuts north of Interstate 80 show excellent exposures of soft-sediment deformation in thinly bedded mudstones of the Mesaverde group. These folds formed when the sediment was still soft and could easily flow. Perhaps an ancient earthquake triggered the flowage.

The Rock Springs uplift is an anticlinal structure that trends north-south through the center of the Greater Green River Basin. The uplift is asymmetric to the west, so its western flank is steeper than the eastern flank; the western flank is also faulted below ground level. The Rock Springs uplift is a miniature version of Wyoming's mountain ranges. It was slowly uplifted during Cretaceous and early Tertiary time. Many oil and gas fields lie along the crest of the uplift.

Older rocks typically crop out in the middle of an anticline, and this is true for the Rock Springs uplift. The upper Cretaceous Baxter shale is exposed in the middle of the uplift and has been eroded to form an oval topographic depression east of Rock Springs. The Mesaverde

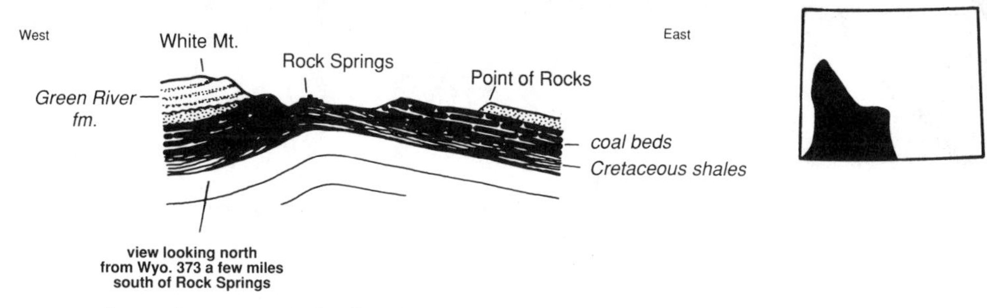

A section across the Rock Springs Uplift.

group forms high ridges of outwardly-dipping sandstone beds around the perimeter of the depression, sometimes called the Baxter Basin, and the Rock Springs uplift.

The steeper west flank of the Rock Springs uplift is traversed between Baxter and Rock Springs. Once again, the interstate highway crosses various members of the west-dipping, upper Cretaceous Mesaverde group, containing white to brown sandstone, gray, sandy shales, and coal beds. Good exposures of these strata are seen in roadcuts along the interstate through Rock Springs.

"Rock Springs" was discovered in 1861 by a Pony Express rider who had to detour to avoid Indians, and discovered the springs that are the town's namesake. A stage coach station was eventually built, and coal mining in the Mesaverde formation by the Union Pacific Railroad became the town's main enterprise for decades. The famed outlaw, Butch Cassidy, worked in a butcher shop in Rock Springs when he was 18, hence the name "Butch." Even today, Rock Springs seems to have an element of the Old West as it booms and ebbs with the flow of mining and oil activity.

Wyoming 789: Creston Junction—Baggs

51 mi./82 km.

This route follows the eastern flank of the Washakie Basin. The Washakie Basin is a large syncline, but it topographically forms a plateau west of the highway because of resistant sedimentary strata of the Eocene Green River and Wasatch formations, capped by sandstone and claystone of the Washakie formation. These Eocene strata were deposited as lake and river sediments between 43 and 50 million years ago. In general, the Green River formation consists of lake sediments, whereas the Wasatch contains river sediments; as the lake expanded and contracted through time, these two deposits became interbedded to form the succession exposed to the west. The

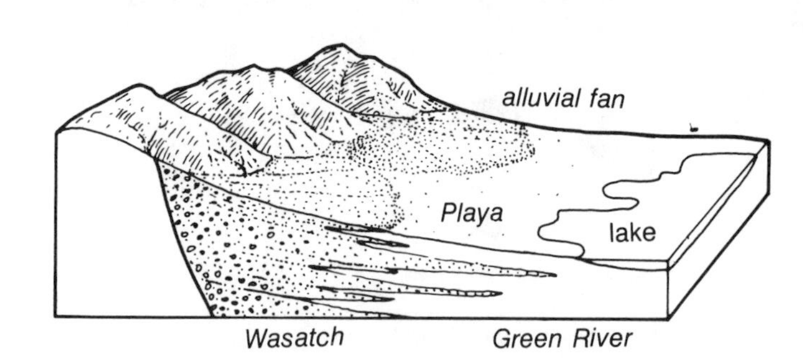

A model of how the Wasatch and Green River formations were deposited in southwest Wyoming. –Adapted from Hamblin (1985)

older Paleocene Fort Union formation, containing brown to gray sandstone, gray shale and thin coal beds, is generally exposed east of the highway.

The southern third of this route follows Muddy Creek, a tributary of the Little Snake River which flows into the Yampa and Green rivers. Baggs was named after the Baggs family who owned a large ranch in the area. Because of its isolation, Baggs was a popular rendezvous spot in the late 1800s for gangs of outlaws, who hid out in the hills and celebrated in town.

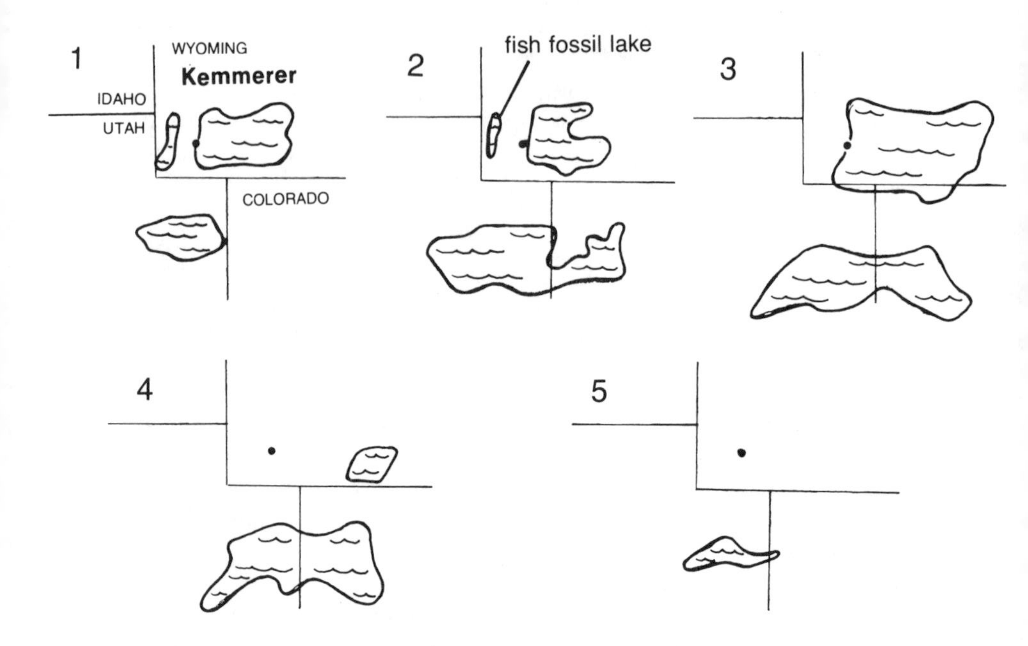

Changes in lake levels throughout the Eocene. One is earliest, five is latest Eocene. –Adapted from McGrew and Casilliano

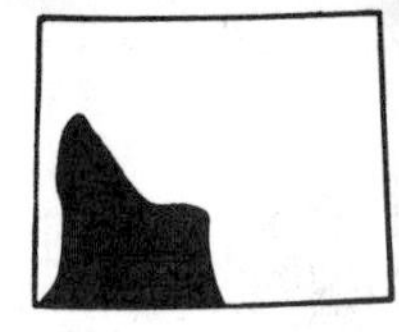

Cliffs of sandstone and claystone of the Green River formation near the town of Green River.

Interstate 80: Rock Springs—Green River—Evanston

14 mi./22 km. 86 mi./138 km.

Interstate 80 between Rock Springs and Green River traverses the steep eastern flank of the Rock Springs uplift and the western margin of the Green River Basin proper. Formations along the west flank of the Rock Springs uplift are the same as those along the east flank, except that the surface outcrops are much narrower because of the steep dip. Within five miles west of Rock Springs, you cross the upper Cretaceous Mesaverde group of brown to white, soft sandstones, gray shale, and coal, the Paleocene Fort Union formation of brown to gray sandstone and black shale, and the inferred trace of the Rock Springs thrust fault.

West of the intersection with US 191, you will see greenish brown sandstones and brown oil shales adjacent to Interstate 80 belonging to the flat-lying Eocene Green River formation. At the town of Green River, twin tunnels cut through this formation. The Green River formation also forms the Palisades, spectacular cliffs on the north side of Interstate 80, as well as the prominent Castle Rock. The origin of the Green River formation is discussed in detail in the section on Fossil Butte; it was deposited approximately 55 to 45 million years ago in a shallow lake that covered most of southwestern Wyoming. The Eocene Wasatch formation was deposited at the same time as shoreline and stream sediments around the lake. The Green River formation can be thought of as a giant lens of lake sediments surrounded by stream sediments of the Wasatch formation.

Like Rock Springs, the town of Green River was a Pony Express and stage station. It greatly expanded during the late 1860s with construction of the Union Pacific Railroad. John Wesley Powell and his party began their historic exploration of the Green and Colorado rivers at the town of Green River in 1869, the first white men to go through the Grand Canyon. Today, visitors can go to the exact spot at Expedition Island in Green River where Powell started his journey. Quoting Powell:

> May 24, 1869 — The good people of Green River City turn out to see us start. We raise our little flag, push the boat from shore, and the swift current carries us down.
>
> Our boats are four in number. Three are built of oak; staunch and firm; double-ribbed, with double stem and stern posts, and further strengthened by bulkheads, dividing each into three compartments. . .The fourth boat is made of pine, very light, but 16 feet in length. . .
>
> We take with us rations deemed necessary to last ten months. . .For scientific work, we have two sextants, four chronometers, a number of barometers, thermometers, compasses, and other instruments.

Powell was one of several geologists who helped to map and open the west in the 1860s and 70s. Other famous geologist-explorers were F.V. Hayden in Wyoming and Montana, Clarence King in Nevada and Utah, and G.M. Wheeler in southern Arizona. Many of the mountains, rivers and canyons were named by these early geologists. One hundred years later, when men went to the Moon, geologists were again among the first to go and explore!

Interstate 80 crosses the Green River about 3 miles west of the town of Green River. Here, the river cut its channel into lake-deposited oil shale and marlstone of the Laney member of the Green River formation, about 45 million years old. The Laney crops out between the Green River and the Blacks Fork River to the west.

The world's largest deposits of trona, the rock that produces the industrial chemical soda ash, are west of Green River. The U.S. Geological Survey estimates that 100 billion tons of trona are in the Wilkins Peak member of the Green River formation, which underlies the area. Trona was precipitated from the waters of Lake Gosiute, an enormous lake that covered over 15,000 square miles of southwest Wyoming during Eocene time, 45 million years ago. Trona is mined

underground by several companies, then refined and shipped by rail throughout the U.S. and exported overseas. It is used extensively in the manufacture of glass, detergents, baking soda, and other chemical products. Two-thirds of the world's supply of soda ash comes from this area, and trona is Wyoming's third largest revenue-generating mineral, following petroleum and coal. Indications of mining operations and refining may be seen from the interstate between Green River and Little America.

Between Little America and Fort Bridger, Interstate 80 crosses outcrops of the Eocene Bridger formation in the Bridger Basin, which is the southern end of the Green River Basin. The Bridger formation contains olive-drab and white sandstones, claystones and conglomerate; it overlies the Eocene Wasatch and Green River formations and was deposited by streams about 45 million years ago. Excellent outcrops of the Bridger formation form a rugged and beautiful badlands topography from Little America to the Bridger Valley Road. The badlands are especially well developed in the Church Buttes area, where Mormon emigrants held a church service on a Sunday in July, 1847.

Fort Bridger was one of the first settlements in Wyoming. It was established as a trading post and blacksmith shop in 1843 by the famous mountain man Jim Bridger, and later served the Mormons, the U.S. Army, and the Pony Express. Wyoming's first newspaper, the "Daily Telegram," was published at Fort Bridger in 1863 and the first schoolhouse in the state was built here in 1866. Fort Bridger State Park is a worthwhile stop for those interested in history.

West of Fort Bridger is a broad, upland plateau on the Eocene Bridger formation. The Uinta Mountains form the high peaks to the

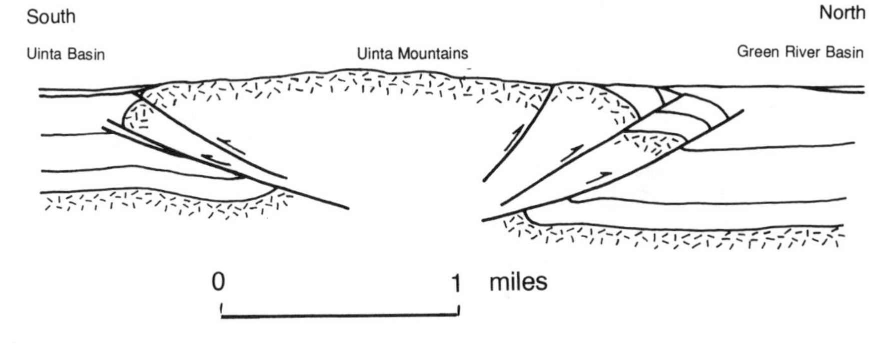

Cross section of Uinta Mountains.

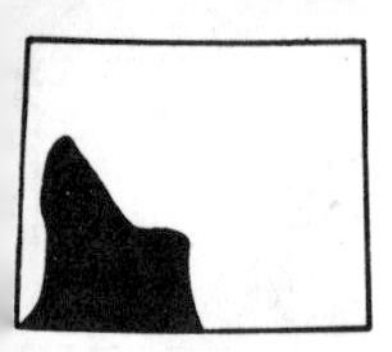

south. The Uintas are composed of sparkling white quartzite that is over 1 billion years old. Like many other Rocky Mountain ranges, the Uintas rose around 55 to 60 million years ago.

Interstate 80 crosses the Darby or "Hogsback" overthrust fault approximately 8 miles southwest of Fort Bridger, in the vicinity of a large parking area adjacent to Interstate 80 at the crest of a hill. This marks the western edge of the Green River Basin, and the beginning of the "overthrust belt" to the west. The Darby is the easternmost overthrust fault in this part of Wyoming's overthrust belt, as discussed in the chapter on the overthrust belt. Rocks in the upper plate were compressed and shoved several miles from the west to their present position, about 60 million years ago. The trace of the fault is near the contact of the Eocene Bridger formation and the Eocene Wasatch formation to the west. The Wasatch formation contains variegated red, brown, and gray mudstones and sandstones with lenses of conglomerate, forming red-colored soil adjacent to the highway.

Hogsback Ridge, also called Oyster Ridge near Evanston, extends north of the interstate just east of the Interstate 80/US 189 intersection. This ridge is composed of the Cretaceous Frontier formation of white to brown sandstone, gray shale, and coal that has been thrust up in the hanging wall of the Darby overthrust fault. The Absaroka overthrust is crossed about 2 miles west of the intersection, but is not well expressed in this area.

The southern end of Fossil Basin is crossed between the Hogsback Ridge and Evanston, with reddish outcrops of the Eocene Wasatch formation on the surface. The Medicine Butte overthrust crops out on the east side of Evanston. Watch for outcrops of the lower Cretaceous Gannett group, which contains red, sandy mudstone, sandstone, and chert-pebble conglomerate.

The Yellow Creek Oil Field overlaps Evanston. This field produces oil, gas, and gas-condensate from the Jurassic Twin Creek limestone and Nugget sandstone, and the Permian Phosphoria formation. Production depths range from 6,300 feet to 12,000 feet. The field was discovered in 1976 and is one of several major oil and gas fields in the Wyoming-Utah overthrust belt that were discovered in the 1970s, causing a major boom in the oil business. The oil-trapping structures are large anticlines in the rocks above the Absaroka overthrust fault.

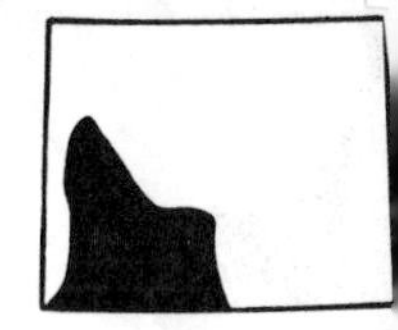

Cliffs of the Eocene Green River formation.

Flaming Gorge Reservoir

Flaming Gorge is a spectacular canyon cut by the Green River as it slices through the north flank of the Uinta Mountains. Geologist John Wesley Powell and his expedition were the first white men to float through the canyon, and they named it "Flaming Gorge" on May 26, 1869, for the bright vermillion rocks forming the canyon walls. Quoting Powell:

> The river is running to the south; the mountains have an easterly and westerly trend athwart its course, yet it glides on in a quiet way as if it thought a mountain range no formidable obstruction. It enters the range by a flaming, brilliant red gorge, that may be seen from the north a score of miles away. The great mass of the mountain ridge through which the gorge is cut is composed of bright vermillion rocks; but they are surmounted by broad bands of mottled buff and gray. . .
>
> This is the first of the canyons we are about to explore — an introductory one to a series made by the river through this range. We name it Flaming Gorge. The cliffs, or walls, we find on measurement to be about 1,200 feet high.

First camp of the John Wesley Powell Expedition, Green River, Wyoming, 1871. — National Archives

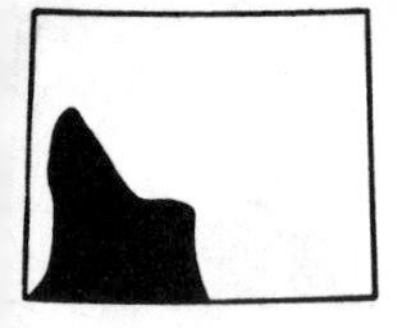

Why did the Green River cut across the Uinta Mountains rather than flow around the east end? Powell thought about this question and eventually developed the theory of "stream superposition," which geologists still find valid today. The first step in stream superposition occurs when uplifted blocks of the crust, like the Uinta Mountains, are buried by eroded sediment. Rivers flow over these buried ranges, not deterred by what lies beneath. In the last few million years, regional uplift caused rivers like the Green to cut canyons into the older, buried mountain range. Therefore, the southerly course of the Green River over the Uinta Mountains was established long ago when the land surface was level with the top of the range.

Flaming Gorge dam was completed in 1964, creating a reservoir that extends over 90 miles north to Green River, and boasts some of the best trout fishing in the west. The canyon walls are red and orange Jurassic and Triassic sandstones which are now under water, and are capped by the Glen Canyon sandstone. Oil shales and marlstones of the Green River formation overlie these strata and form the surface outcrops around the reservoir.

Two routes may be followed from Interstate 80 to Flaming Gorge. Wyoming 530 follows the west side of the reservoir, whereas US 191 follows the east side and gives access to the spectacular Firehole region. The two Wyoming highways are connected by Utah 44 and Utah 260, offering a 160-mile loop drive around Flaming Gorge.

US 191: Rock Springs—Pinedale—The Rim

116 mi./186 km.

The road from Rock Springs to Pinedale becomes progressively more scenic as it approaches the southern flank of the Wind River Mountains. You traverse the Green River Basin proper, with the Rock Springs uplift and Great Divide Basin to the east, and the Wyoming overthrust belt, forming snow-capped peaks, to the west. This section of road seems "typically Wyoming" in that the landscape has a dual and contradictory character to it — the vast and empty basins are framed on the horizon by snowy mountain peaks.

Subsummit surface on the flanks of the ranges, to which sediment was filled during early Tertiary time. Since then, a lot of material has been removed by the erosive power of the Green River!

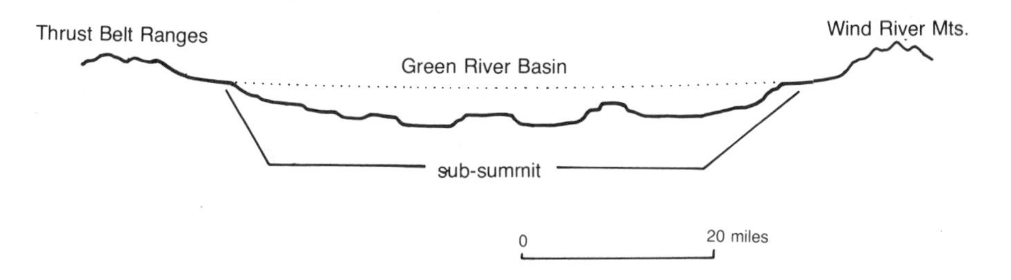

Geologically, most of this road is on two bedrock formations: 1) the Laney member of the Eocene Green River formation, and 2) the New Fork Tongue of the Eocene Wasatch formation. These rocks are described below.

For ten miles north of Rock Springs, US 191 diagonally cuts across the western flank of the Rock Springs uplift. White and brown sandstone, gray shale, and coal on the east side of the road belong to the upper Cretaceous Mesaverde group, whereas White Mountain, to the west across Killpecker Creek Valley, is composed of the Laney member of the Eocene Green River formation. Pilot Butte, the highest point on White Mountain at 7,932 feet, was a famous landmark for the early pioneers. The road crosses into the Laney member about 12 miles north of Rock Springs where the road turns west and climbs up White Mountain; it is exposed for the next 50 miles to the north. The Laney member contains thinly laminated, brownish oil shale and marlstone, a clayey limestone or calcareous mudstone; it is about 45 million years old, and was deposited as fine-grained sediment at the bottom of the Green River Lake which is discussed in the section on Fossil Butte.

The Killpecker Sand Dunes start a few miles east of Eden. The prevailing wind blows from west to east — stop your car and test this for yourself! — and funnel through a gap between the Leucite Hills. The Leucite Hills are Pleistocene volcanic flows and necks containing

The flat subsummit erosional surface at the north end of the Wind River Range as seen from Highway 191 in the northern Green River basin.

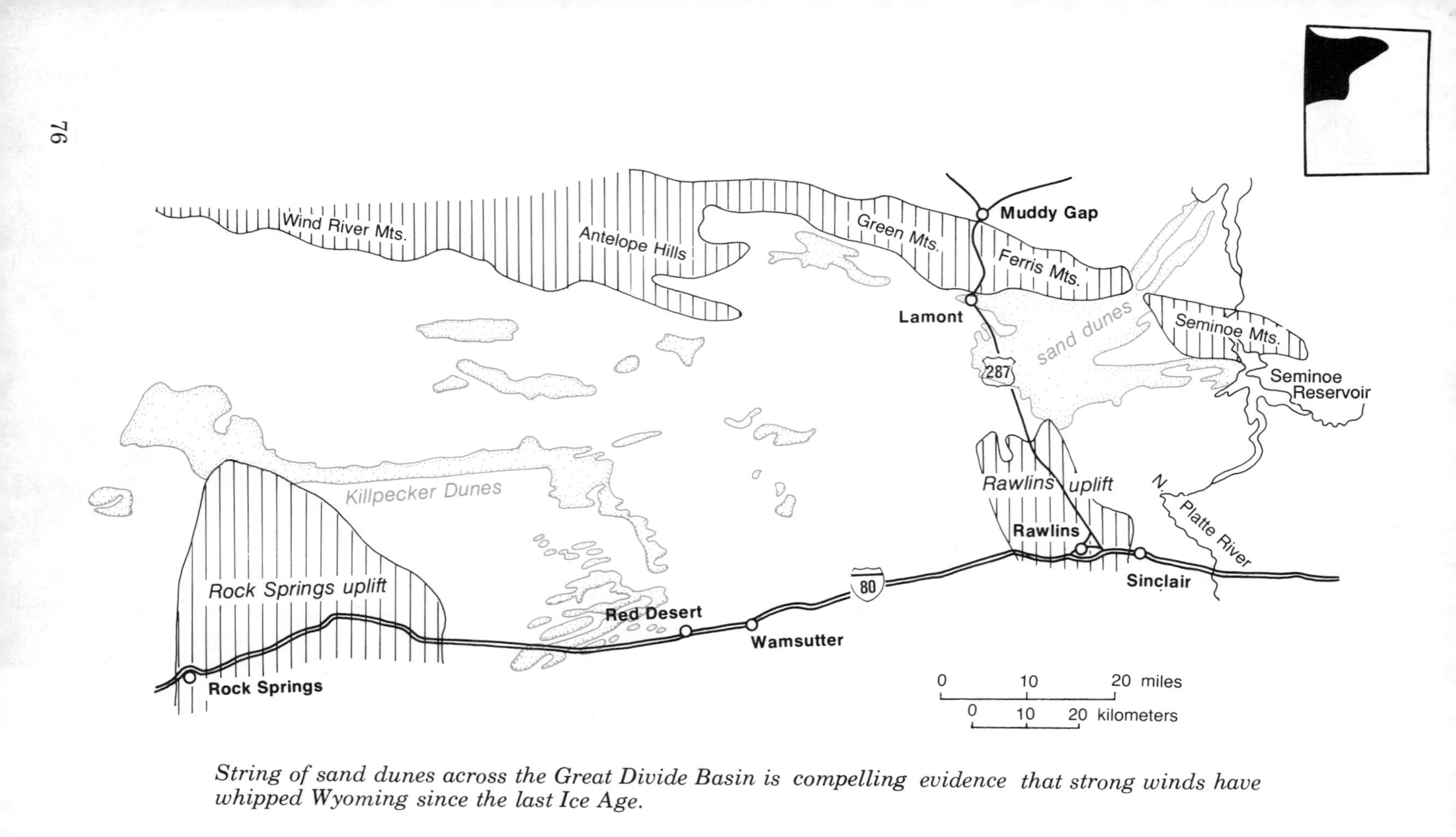

String of sand dunes across the Great Divide Basin is compelling evidence that strong winds have whipped Wyoming since the last Ice Age.

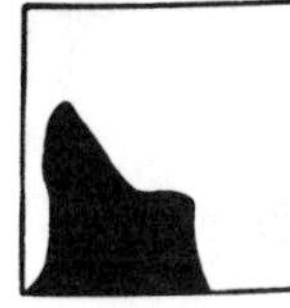

leucite, a mineral found in some potassium-rich volcanic rocks. The winds blew loose sand that was piled against the Leucite Hills through the wind gap and carried it 50 miles downwind to the east into the central Great Divide Basin, forming enormous sand dunes up to 150 feet high, known as the Killpecker Dune Field. The name comes from Killpecker Creek, colorfully named by the U.S. Calvalry in the 1860s because of the mineral-rich water's effect on the troops.

The Sublette Cut-off of the Oregon Trail is crossed north of Big Sandy Reservoir. Sometimes called the "Dry Drive" because of the scarcity of water, this route crossed directly to the Bear River in Idaho and bypassed Fort Bridger to the southwest. It is ghostly to think of the thousands of pioneers who passed here on their way to a new life in the West in the mid-1800s. Today nature has reclaimed almost all evidence of their passing.

The northern fourth of the road between Rock Springs and Pinedale traverses the New Fork Tongue of the Eocene Wasatch formation. This unit consists of dull, variegated red and green mudstone, brown sandstone, and thin beds of limestone. US 191 parallels the New Fork River between Boulder and Pinedale.

At Pinedale, take time to see Fremont Lake, a few miles northeast of town, at the base of the majestic Wind River Mountains. Fremont Lake is a natural lake 600 feet deep, with 22 miles of shoreline; it formed behind a dam of glacial debris, a moraine, that was pushed out of the mountains during the last major glaciation. This glaciation is called the "Pinedale Glaciation" because of the well developed morainal ridges around Fremont Lake. Three major glaciations have occurred in the Rocky Mountains over the past 200,000 years or so. The Pinedale glacial advance was from 70,000 to 15,000 years ago;

Fremont Lakes, northeast of Pinedale.

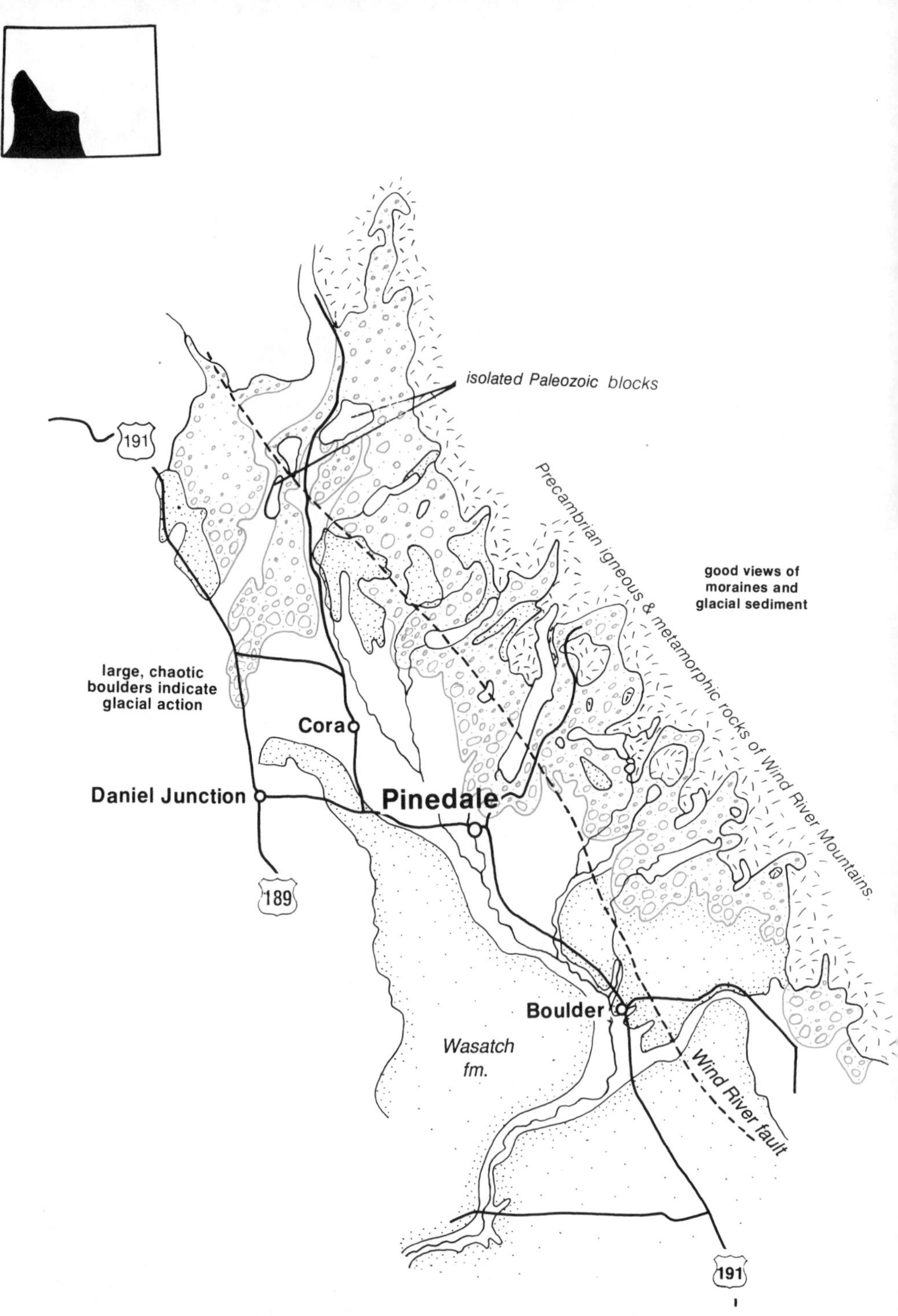

Glacial deposits of the Green River lakes area.

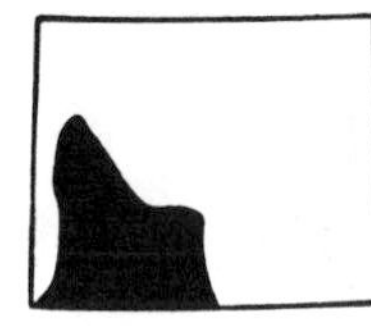

Fremont Lake northeast of Pinedale.

the next oldest was the Bull Lake glacial advance from 200,000 to 130,000 years ago; and the oldest was the pre-Bull Lake or Buffalo glacial advance more than 200,000 years ago. Snow and ice accumulated in high mountain valleys above timberline and eventually formed a moving glacier, which flowed downhill and carved spectacular U-shaped valleys; it pushed eroded rock and dirt debris in front like a bulldozer. The larger glaciers, like the Fremont Canyon and Green River Lakes glaciers, carved their way to the base of the mountains and moved over the floor of the Green River Basin. When the glaciers melted back into the high canyons, they left natural moraine dams across the paths of meltwater streams, and then glacially carved valleys filled with the lakes we see today. Other examples of moraine-dammed lakes in Wyoming include Jackson, Jenny and other lakes at the east base of the Tetons, and Bull Lake on the north side of the Wind River Mountains — the type or standard location of the Bull Lake glaciation.

To the north and east of Pinedale rise the high peaks of the Wind

Boulder moraine, Fremont Lake, northeast of Pinedale.

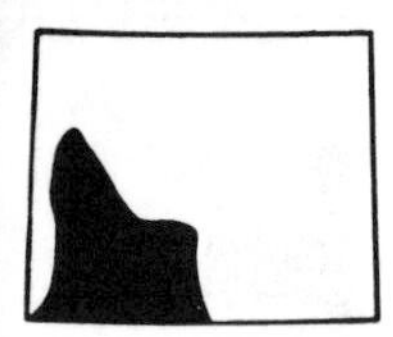

River Range, including the highest point in the state, Gannett Peak at 13,785 feet. Wyoming 352, about 6 miles west of Pinedale, offers access to the Green River Lakes, headwaters of the Green River. The road is paved for some distance, then becomes gravel, so make inquiries about its condition before proceeding. The high peaks above the lakes are home to the Dinwoody Glaciers, the largest accumulation of active, alpine glaciers in the continental U.S.

The valley of the Green River west of Pinedale was the site of the "Upper Green River Rendezvous," where trappers, traders and Indians gathered during six summers between 1833-1840 to swap beaver pelts, or "plews," for supplies and kick up their heels. These gatherings were the highlight of the trapper's year. In solitude, he trapped beaver in the cold mountain streams all winter long, preparing for summer when he could trade his plews for supplies that had been brought from St. Louis. Popular sporting events at the rendezvous included drinking, fighting, horse racing, and wild carousing. The Trappers Point Vista west of Pinedale offers a scenic view of the rendezvous site and the Green River Basin. The sparsely vegetated

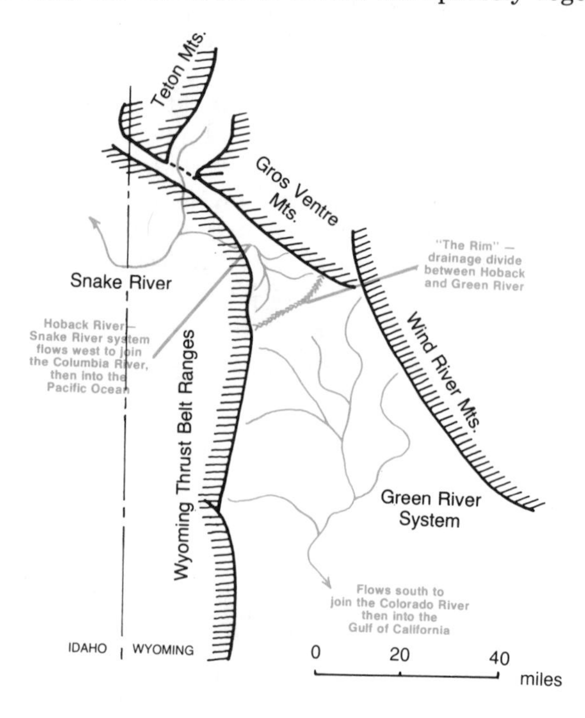

Separate river systems drain one large structural basin in western Wyoming. –Adapted from Dorr, Spearing and Steidtmann (1977, p. 65)

hills above the river are composed of mudstones and sandstones of the Eocene Wasatch formation. The Hoback Range, in the overthrust belt, rises on the western horizon.

North of Pinedale to "The Rim" of the Hoback Basin, US 191 crosses red-brown-gray mudstones, sandstones, and conglomerate lenses of the Wasatch formation. Approximately 2 miles south of the Hoback Rim, the highway crosses "Bull Lake" age glacial moraines deposited more than 130,000 years ago when ice of the Green River glacier extended 5 miles over the basin floor from the mountain front. The Rim is a small pass (elevation 7,921 feet) that forms a topographic divide between the Green-Colorado drainage basin and the Hoback-Snake-Columbia drainage basin. The summit of The Rim is held up by relatively resistant sandstones and mudstones of the Eocene Wasatch and Pass Peak formations. The road section through the Hoback Basin and Hoback Canyon, to Jackson Hole, is covered in the chapter on the overthrust belt.

US 189: Evanston—Kemmerer—Daniel

51 mi./82 km. 84 mi./134 km.

US 189 intersects I-80 about 13 miles east of Evanston. From this point north to Kemmerer, US 189 follows a broad valley cut by Albert Creek into gray and tan siltstones and shales of the Cretaceous Hilliard formation. The prominent ridge flanking the highway to the east is the "Hogsback," also called Oyster Ridge at Kemmerer, for fossilized oysters in the upper strata; it is composed of white to brown, resistant sandstones of the Cretaceous Frontier formation. Both the Hilliard and Frontier formations were deposited in marine waters of the "Western Interior Seaway" that covered most of the mid-continent throughout the Cretaceous Period, before the Rockies were uplifted. Then, in early Cenozoic time, compressional forces uplifted, folded, and moved the sedimentary layers several miles eastward. The Hogsback or Oyster Ridge, is formed by the rocks in the hanging wall of the Darby thrust fault, which surfaces a few miles east of the Hogsback.

Kemmerer is a major coal mining center in this part of southwestern Wyoming. Most of the coal mining is in the upper Cretaceous

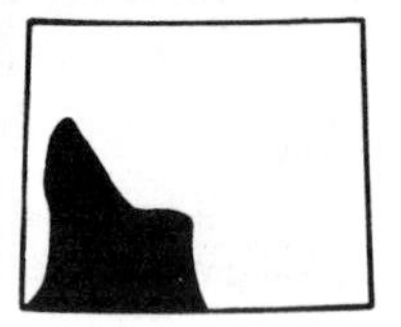

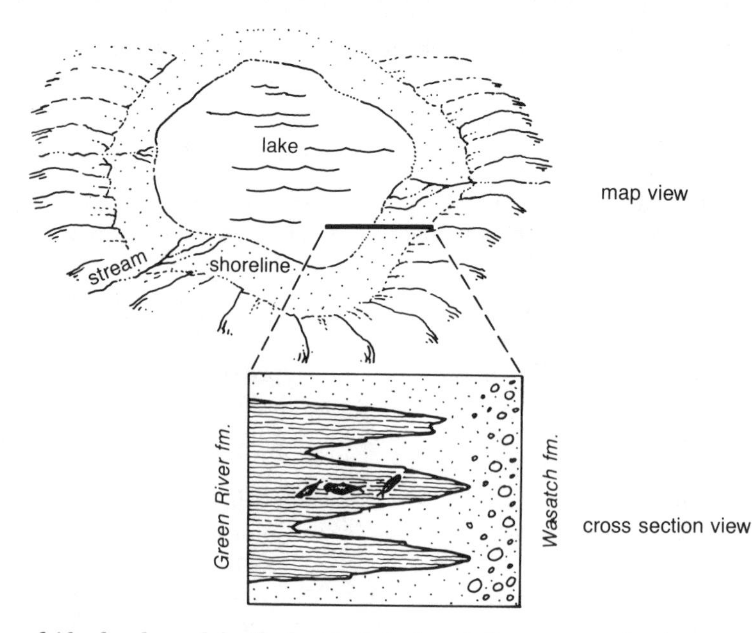

Shoreline shifts back and forth due to lake level changes producing interfingering of Green River and Wasatch formations. –Adapted from Oriel (1962)

Frontier and Adaville formations. The massive Kemmerer Coal Company plant is visible from the highway just south of Diamondville. The first J.C. Penney Company store in the U.S. was opened in downtown Kemmerer in 1902, and is still open for business!

Fossil Butte National Monument, ten miles west of Kemmerer on US 30, contains a remarkable concentration of fossilized, fresh-water fish, as well as other aquatic animals, insects, a few mammals, and plants. The 1,000-foot-high Fossil Butte is composed of red, purple, yellow, and gray sandstone and claystone beds of the Wasatch formation at the base, whereas the buff-to-white colored claystones and marlstones (lime-rich claystones) of the Green River formation form the steep cliffs along the top. The Wasatch formation was deposited mainly by streams, whereas the Green River is a lake deposit. During Eocene times, 45 to 50 million years ago, when these sediments were laid down, this area was a humid, forested, sub-tropical environment! Some of the fossil plant remains include palm tree fronds.

North of Kemmerer, US 189 follows west-dipping sands and shales of the Frontier formation along Oyster Ridge for about five miles, then turns sharply east. Fossilized marine oyster shells abound in sandstones along the top of the Frontier formation. East of Oyster Ridge, the highway traverses Eocene lake and stream deposits that cover lower Cretaceous and Jurassic rocks in the rocks above the Darby thrust fault. The road crosses the trace of the Darby thrust on a

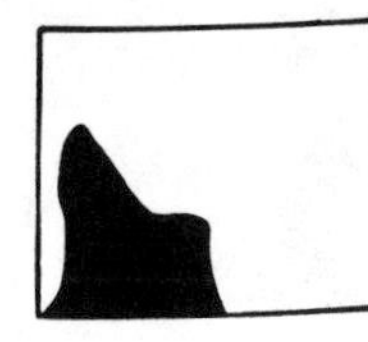

large hill about halfway between Kemmerer and Fontenelle Reservoir, about 13 miles north of Kemmerer. From here to Fontenelle Reservoir, US 189 traverses a broad upland plateau on lake-deposited, fine-grained oil shale and mudstones of the Eocene Green River formation.

Fontenelle Reservoir on the Green River is named after Lucien Fontenelle, an outfitter for the American Fur Company in the 1830s.

The bluffs surrounding Fontenelle Reservoir are brown to buff, fine-grained lake sediments of the Eocene Green River formation, interbedded with stream-deposited sediments of the Eocene Wasatch formation. During Eocene time, a large lake, called Lake Gosiute, covered southwestern Wyoming and northeastern Utah, where it is called Lake Uinta. The fine-grained, light-colored sediments, including oil shale, of the Green River formation were deposited within these lakes, while stream and floodplain sediments of the Wasatch formation were deposited around the lakes. As the lakes shrank and expanded seasonally throughout the Eocene, the Green River and Wasatch formations became interbedded, or as geologists say, intertongued.

About 3 miles south of La Barge is Names Hill, a spot along the Oregon Trail where emigrants stopped to carve their names into soft marlstones of the Green River formation; the inscription "James Bridger, 1844" is still legible! To the southeast is the site of Mormon Ferry, built in 1847 by Mormons under the leadership of Brigham Young.

Between La Barge and Big Piney, the La Barge and Chappo members of the Eocene Wasatch formation crop out adjacent to the road and form the bluffs to the east above the river. They are overlain by the New Fork Tongue of the Wasatch. The La Barge and Chappo members are red-gray-brown variegated mudstones and conglomerates with yellow sandstones, whereas the New Fork Tongue units were deposited by streams that flowed east from the overthrust belt across the newly formed Green River Basin around 45 to 50 million years ago. The high peaks of the overthrust belt still rise to the west. The rugged crest of the Wind River Range appears in the northeast.

The La Barge — Big Piney area is an important oil and gas producing area. When oil was first found here in the 1920s, the La Barge oil field was called "Tulsa" after the prosperous, oil-rich city in Oklahoma, but after the boom ended in 1935 the field and town were

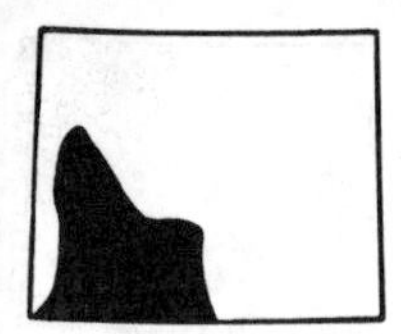

renamed after nearby La Barge Creek, named for an early 1800s river pilot. Several fields are in the hills around La Barge and west of Big Piney, producing oil and natural gas from lower Tertiary, Cretaceous, and Jurassic sandstones. Petroleum accumulated in this area because strata were broadly folded into an anticline called the Moxa arch, a large flexure in the Precambrian basement that uplifted overlying strata. The Moxa arch trends north-south through the western Green River Basin. Many oil and gas fields lie along its length.

Between Big Piney and Daniel, US 189 crosses several tributaries of the Green River, such as Muddy Creek and Cottonwood Creek. These creeks cut into the La Barge and Chappo members of the Wasatch formation. These Eocene strata are overlain by Quaternary gravels derived from the adjacent mountains.

Picturesque badlands just south of Cottonwood Creek expose variegated red and tan claystones of the Wasatch formation. The Hoback Range rises above the plains due west, forming the western margin of the Green River Basin.

Daniel was a popular site for fur traders' rendezvous during the summers of 1833 through 1840.

Major John Wesley Powell and his daring crew at Green River, Wyoming as they prepare to depart on their exploratory journey down the Colorado River. May 24, 1869. —credit National Archives

III
Northeastern Wyoming — The Powder River Basin

INTRODUCTION

"Powder River: Let 'er Buck!!" This was the cry made famous by thirsty cowboys after a long, dusty cattle drive from Riverton to Casper. It's been said that Powder River cowboys never try to keep their horses from pitching, they just "Let 'er Buck!!" The Powder River Basin is Wyoming's corner of the High Plains. It has evolved from Indian country to cowboy country, and now is home to oil drillers and coal miners. Situated between the Black Hills and the Bighorn Mountains, the Powder River Basin contains a wealth of energy resources beneath wind-blown grasses and buried buffalo bones.

The basin was named after the Powder River for the dark, fine-grained soil, resembling gun powder, that occurs along its banks. The Powder River, "a mile wide and an inch deep," is said to contain "water too thick to drink and too thin to plow." The Indians would rub dirt between their fingers, then let it fall as a fine dust to signify the Powder River. The river flows almost due north through the center of the basin from its headwaters in the southwest corner, and eventually joins the Yellowstone River east of Miles City, Montana.

Major tributaries to the Powder River include Crazy Woman Creek and Clear Creek, which drain the eastern slopes of the Bighorn Mountains. The Little Powder River and Belle Fourche River (pronounced Bell Foosh) drain the northeast

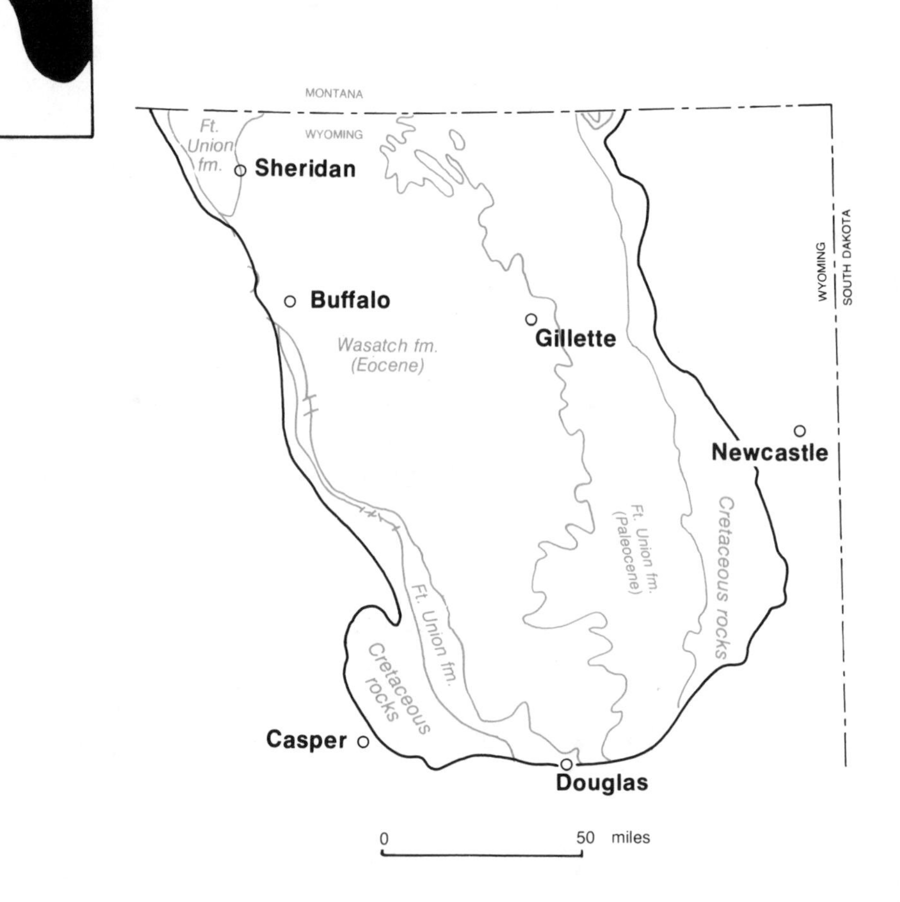

Geologic map of the Powder River Basin. –Adapted from Specht and Bryant (1979)

corner of the basin, while the North Platte and Cheyenne rivers drain to the south.

The Belle Fourche River, or "Beautiful River," was named by French trappers sometime before 1870. Indians called the Belle Fourche the "Bear Lodge River". The Belle Fourche flows over the lowest spot in Wyoming, 3,100 feet, as it exits the State; the highest spot is 13,785 feet at the summit of Gannett Peak in the Wind River Mountains.

The Powder River Basin contains the Thunder Basin National Grassland, which covers approximately 2 million acres. This semi-arid land, properly managed, provides valuable grazing for cattle, sheep, bison, and one of Wyoming's largest herds of pronghorn antelope.

Indians versus Gold

The geological framework of the Powder River Basin, and the gold of adjacent mountains, played a major role in the history of the High Plains. Initially, the Powder River Basin was a prized hunting ground for the Sioux, Cheyenne, and Arapahoe tribes. It was the transient home of millions of buffalo — one herd could cover hundreds of square miles and blacken the horizon with their shaggy coats. The following quote summarizes why the Indians valued the Powder River Basin and the country around it so much:

> From the Indian's standpoint, the country was all that could be desired - plenty of shelter, mild winters, sufficient food for their ponies, and an abundance of game in the broken country surrounding them for their own subsistence. It was an Indian paradise.
>
> T.J. Gatchell, from Larson's *History of Wyoming*

The idyllic lifestyle of the Plains Indians began to deteriorate by the mid-1800s due to the inevitable incursion of the white man. Sioux Indians, under the great Chief Red Cloud, won several battles against the U.S. Cavalry, which led to the establishment in 1868 of the Great Sioux Reservation, extending from the Missouri River to the Black Hills. The Sioux wanted their traditional hunting grounds in the Powder River Basin, but the treaty was deliberately vague on sovereignty. Chief Red Cloud took his band to the reservation in hopes of saving lives, but younger, militant chiefs, like Crazy Horse of the Ogallala Sioux and Sitting Bull of the Hunkpapa Sioux, stayed by the Powder River. When General George Armstrong Custer reported "gold at the roots of the grass" during his 1874 expedition, prospectors flooded the Black Hills. The Sioux protested and fought to retain their land, with Crazy Horse and Sitting Bull leading the resistance. Then, on a gentle June day in 1876, Custer's 7th Cavalry of little more than 200 men, blindly charged into more than four thousand Sioux Indians camped along the Little Big Horn River in Montana. Even though the ensuing battle became Custer's Last Stand, it ironically marked the end of the Indian resistance, for the Indians were bitterly defeated the next winter by General Crook.

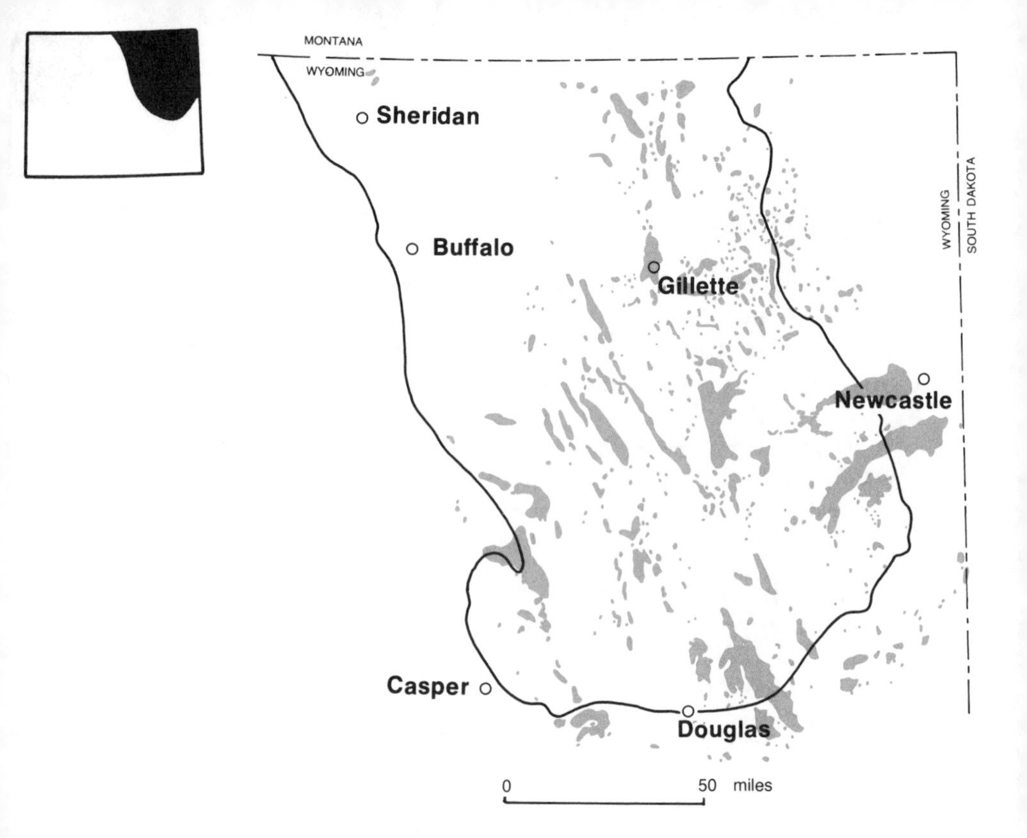

Oil and gas fields in the Powder River Basin. –Adapted from the Geological Survey of Wyoming's 1980 Oil and Gas Fields Map of Wyoming.

Oil and coal are now the targets of modern plains hunters, but the Powder River Basin is still home to one of the largest buffalo herds in North America.

POWDER RIVER ENERGY RESOURCES

Petroleum

The Powder River Basin has produced oil and gas for decades. The most famous oil fields in the basin, perhaps in the country, are Salt Creek and Teapot Dome. Located on the southwestern shoulder of the basin, 40 miles north of Casper, these two oil fields helped shape the history of Wyoming and were the focus of a national political scandal.

Other major concentrations of oil in the Powder River Basin occur along the central and eastern parts of the basin. Production comes from a variety of lower and upper Cretaceous strata, as well as from upper Paleozoic strata in the northeastern part of the basin.

Coal

Coal resources in the Powder River Basin are enormous! More than 12,000 square miles overlie coal-bearing strata. The coal reserves beneath Campbell County alone could supply the Nation's needs for the next 200 years. Powder River coal is highly valued by utilities because of its low sulfur content and low concentrations of toxic elements. Coal mining is economic if coal beds are close to the surface — this is true throughout the basin. The thickest and most important coal beds are in the upper part of the Paleocene Fort Union formation, the Tongue River member. The Wyodak-Anderson coal bed, near Gillette, is Tongue River coal that ranges up to 125 feet in thickness and extends for more than 100 miles along the eastern side of the basin. Also, the Eocene Wasatch formation contains important

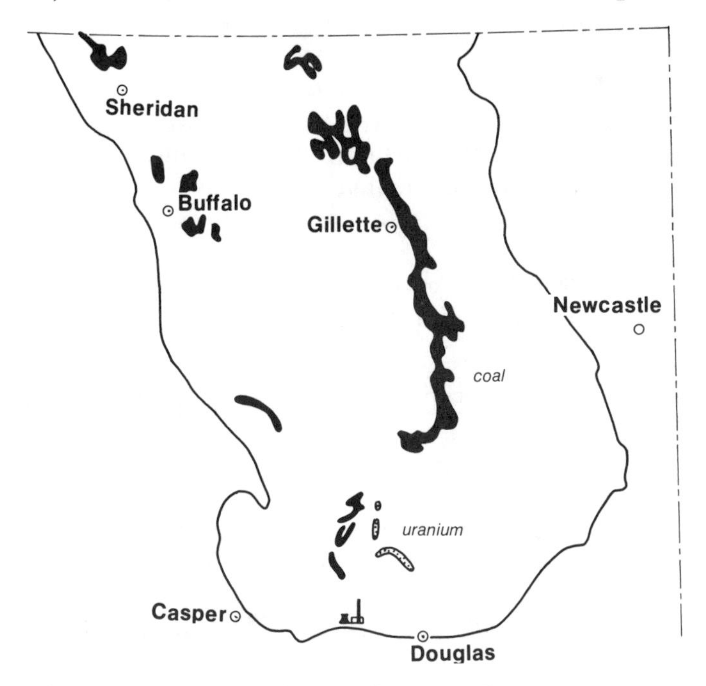

Coal and uranium deposits in the Powder River Basin. –Adapted from Specht and Bryant (1979)

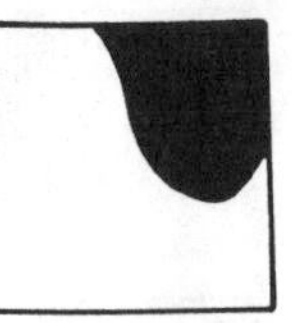

coal beds near Lake De Smet on the west side of the basin, where the Healy coal is over 200 feet thick. In general, Fort Union coals are more important in the eastern part of the basin, while Wasatch coals are more important in the central and western parts. Coal production has boomed since the 1970s and Powder River coal now fires power plants throughout the country.

Many of the coal beds in the basin are exposed at the surface. In the past millennia, lightning struck some of these outcrops and ignited the coal, causing slow burning that lasted for years, decades, perhaps even hundreds of years. As a result, the shales and sandstones surrounding the coal were baked, or thermally metamorphosed, and melted to form a rough, reddish-orange, frothy slag called "clinker," a distinctive trademark of the Powder River Basin. It can be seen between Sheridan and Buffalo, and between Wright and Gillette. Clinker has been quarried and used extensively as road gravel. Watch for the red highways!

Uranium

The Powder River Basin also contains some of Wyoming's largest uranium deposits. Uranium was first identified in Wyoming in 1918 at the Silver Cliff mine near Lusk. In 1951, J.D. Love of the U.S. Geological Survey discovered uranium in the Pumpkin Buttes area, ushering in commercial exploration and development. The uranium was deposited from ground water in tiny pore spaces between sand grains in the Wasatch and upper Fort Union formations.

SURROUNDING MOUNTAINS

Black Hills

The Black Hills are an oval uplift, a dome, that straddles the border between South Dakota and Wyoming. The hills are approximately 80 miles long by 40 miles wide, as measured from the "red racetrack" of soft Triassic shales that crop out

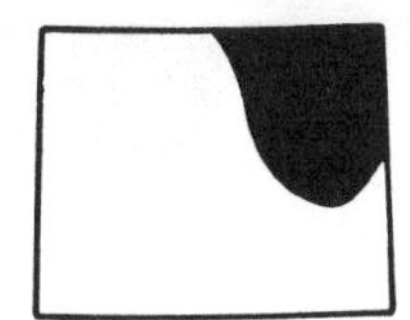

around the perimeter. Because the hills are a domal uplift, the oldest rocks are exposed in the center, flanked on all sides by progressively younger strata outward. The rocks in the center are Precambrian "basement," mostly granitic rocks.

The northern Black Hills contain large gold deposits at Lead, South Dakota. The Homestake Mine was discovered in 1876, with the demise of Custer on the Little Big Horn. Gold is concentrated in folded and metamorphosed Precambrian sedimentary rocks, particularly carbonate rocks of the Homestake formation. Apparently gold mineralization occured more than once during the history of these rocks, including the period of intrusion of Eocene igneous rocks related to Devils Tower.

Bighorn Mountains

The Bighorn Mountains frame the western margin of the Powder River Basin. They form a large arc of uplifted Precambrian "basement" rocks that runs from the Pryor Mountains in Montana to the east end of the Bridger Mountains in central Wyoming. Cloud Peak, 13,165 feet, is the highest point along the glacier-carved spine of the range. The Cloud Peak Wilderness Area established in 1974, encompasses 137,000 acres of alpine land above 8,500 feet.

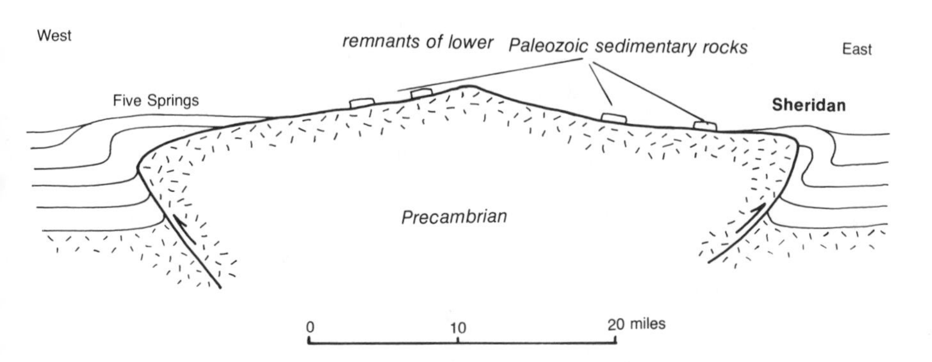

Cross section of Bighorn Mountains.

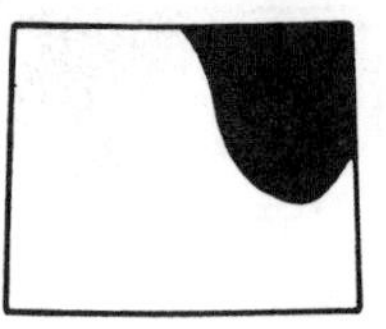

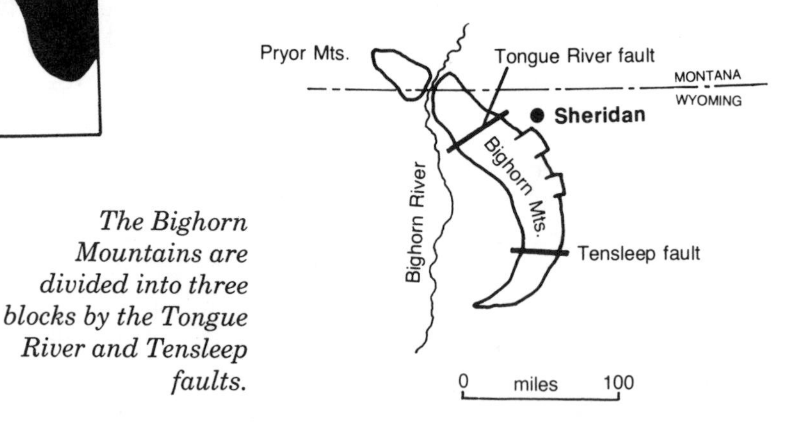

The Bighorn Mountains are divided into three blocks by the Tongue River and Tensleep faults.

About 60 million years ago, the area that is now the Bighorns began to bow upward, due to compression of the crust, while basins to the east and west began to sag. This continued for several million years, causing the crust to rupture, creating thrust faults on the east and west side of the range. The Paleozoic and Mesozoic sedimentary rock that once covered the top of the range were mostly eroded away by streams during uplift, but are preserved along the flanks of the range where they steeply dip into adjacent basins. The Bighorn Range differs from other Wyoming ranges that are bound on only one side by a thrust fault, like the Wind River and Owl Creek ranges. For example, the Wind River Range is bound by a major thrust fault on its southwest flank, while the back side, or northeast flank, dips gently into the Wind River Basin.

The Bighorn Range may be subdivided into three structural segments. The northern segment is north of Alternate 14, which crosses the northern part of the range. This segment was shoved to the southwest over the Bighorn Basin by the Five Springs thrust fault, which crops out along the western base of the range where Alternate 14 leaves the mountains. The central segment extends from US 14 south to US 16; this segment was shoved eastward over the Powder River Basin by thrust faults that crop out along the eastern base of the range. The southern segment extends south of US 16 to the east end of the Bridger Mountains. This segment is bounded along its western flank by the Bigtrails fault, a high-angle reverse fault. Thus, during the Laramide orogeny the Bighorns were shoved west at the north and south ends, and east in the central part of the range, as the crust was squeezed and shortened.

Casper Area

Casper has a long and fascinating history. The area was occupied for thousands of years by bands of paleo-Indian hunters, followed by Plains Indian cultures of the 1700 and 1800s. After the Louisiana Purchase of 1804, exploration and fur trade brought many adventurers to central Wyoming. The North Platte River served as a natural route for western migration. The first commerical ferry in the Rocky Mountains was started in 1847 by Brigham Young at the future site of Fort Caspar. As emigrant traffic increased, Indian hostilities increased, and the U.S. Army moved into the area. The Battle of Platte Bridge Station in 1865 pitted a thousand warriors against a small group led by Lt. Caspar Collins. The Indians were victorious, but Collins fought and died so heroically, the Platte Bridge Station was renamed Fort Caspar in his honor (Fort Collins in Colorado had already been named for his father). The spelling was changed to Casper in 1888 by railroad officials.

The first oil well near Casper was drilled in 1888, but it "came up dry." The initial oil boom began in 1908 with the first completed well at Salt Creek, 40 miles north. It was a spectacular gusher and word of its discovery spread across the country. Thousands of people rushed to Wyoming between 1912 and 1918, and Casper's population grew from 2,000 to 20,000. In the years since, Casper has been the regional center for many oil and mining boom-and-bust cycles, and is today a hub for energy exploration and production in the central Rockies.

Casper lies at the northern base of Casper Mountain, a large anticline with Precambrian rocks in the center, at the north end of the Laramie Range. Casper Mountain was uplifted and thrust to the north over the south margin of the Powder River Basin during the Laramide orogeny. The Casper Mountain thrust fault crops out along the northern base of the mountain and dips south between 30 and 40 degrees beneath the uplift. Muddy Mountain, south of Casper Mountain, is composed of red Mesozoic shales and underlying Paleozoic strata on the gentle, backside of Casper Mountain. The top of Casper Mountain is composed of ancient Precambrian metamorphic and

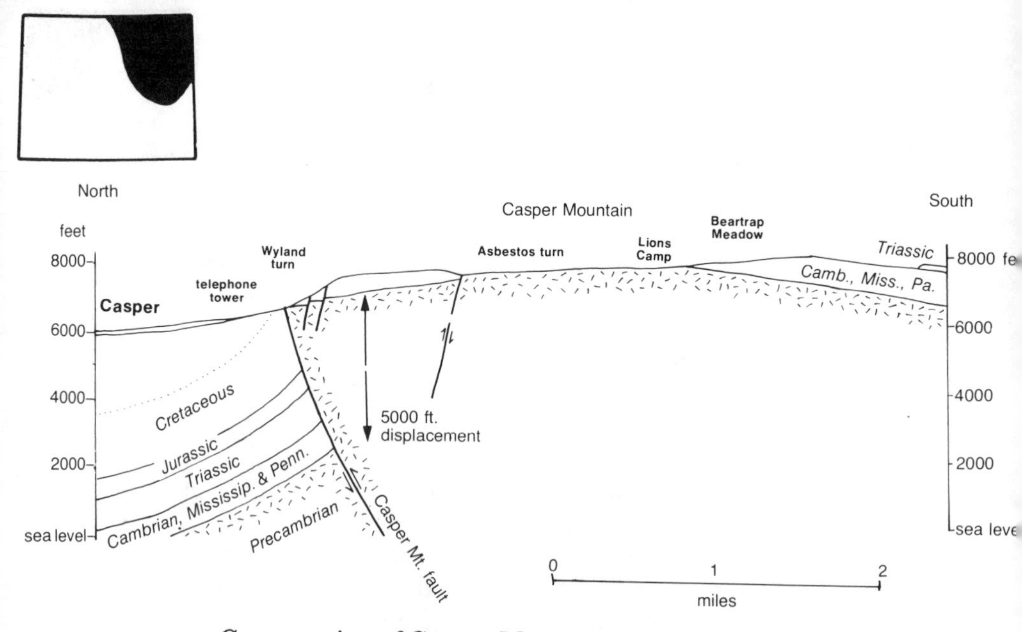

Cross section of Casper Mountain. –Adapted from Knittel (1978)

igneous rocks, flanked by gently north-dipping and south-dipping Paleozoic strata.

Highway 251
Casper Mountain Road

Wyoming 251, the Casper Mountain Road, crosses Casper Mountain and joins Wyoming 487 to the south. As you drive south toward Casper Mountain, you travel over geologically recent terrace gravels that were deposited over black shales of the Cretaceous Cody formation. At the base of the mountain, the road grade increases markedly and begins to switch-back up the north face of the mountain. The Casper Mountain thrust fault is crossed between the first and second hairpin turns of the road at the base of the steep incline. After crossing the thrust fault, the road climbs through Precambrian metamorphic rocks, the Cambrian Flathead sandstone, the Mississippian Madison limestone, and near the top, the Pennsylvanian Casper (Tensleep) sandstone. You drive through these units again on the backside (south side) of Casper Mountain toward Muddy Mountain. The Precambrian metamorphic and igneous rocks on the top of Casper Mountain have yielded small deposits of asbestos and feldspar; there was even a short-lived copper boom around the turn of the century.

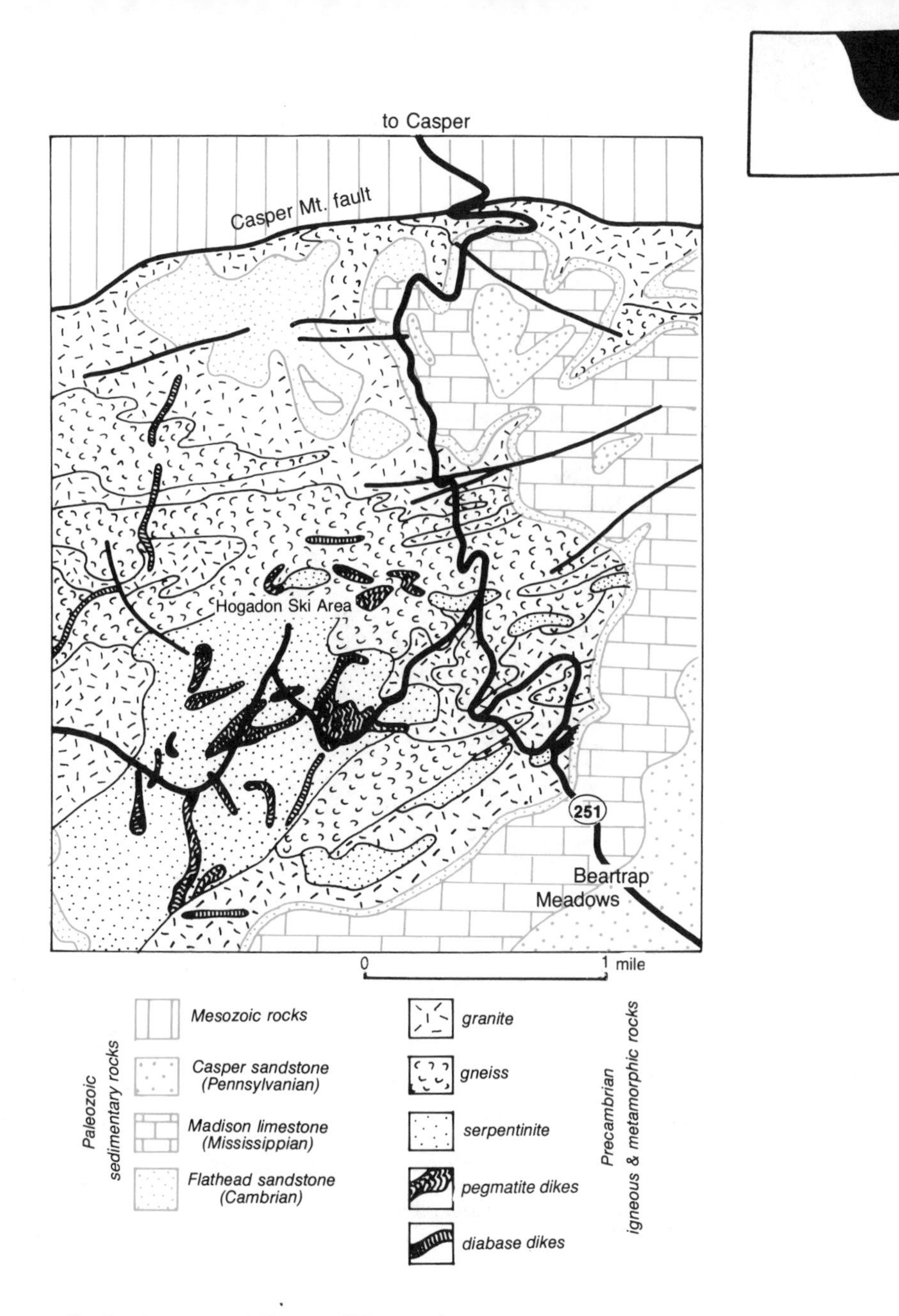

Geologic map of Casper Mountain. –Adapted from Burford et al. (1979)

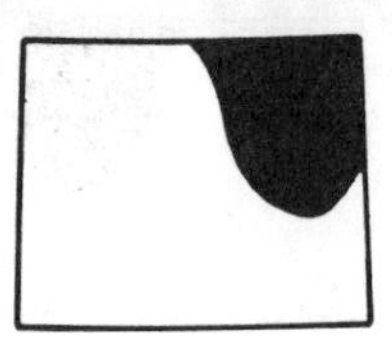

EASTERN POWDER RIVER BASIN

Interstate 90
Spearfish, S.D.—Gillette
95 mi./152 km.

Interstate 90 enters the northeastern corner of Wyoming at the north end of the Black Hills. This is a beautiful section of highway, noted for its red rocks, pine covered hills, and bold outcrops of igneous rocks like Devils Tower. The Black Hills are a domal uplift with ancient Precambrian rocks in the center. Younger Paleozoic and

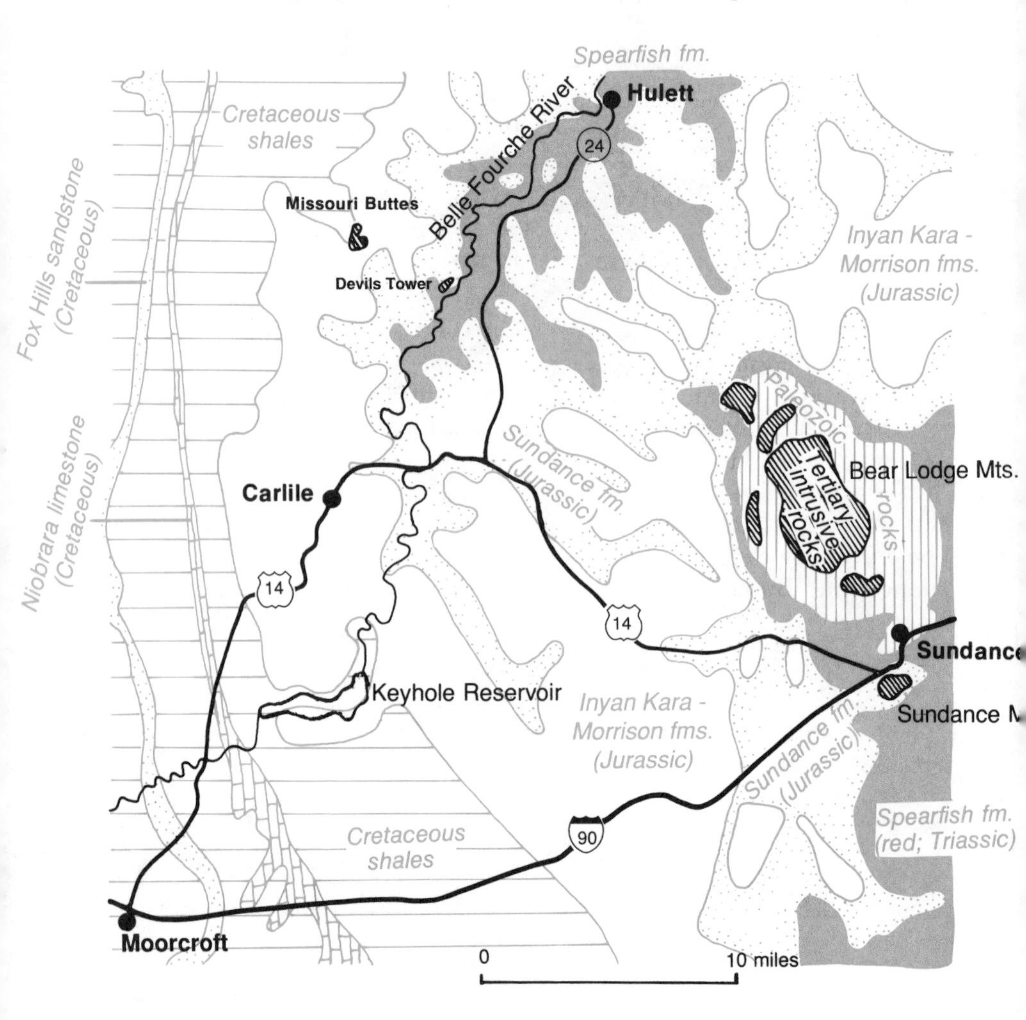

Geologic map of Devils Tower and Sundance area, northeast Wyoming.

Diapir at Sundance, Highway I-90

Mesozoic sedimentary strata dip away from the uplift on all flanks. One formation in particular, the Triassic Spearfish formation, is composed of soft red shale that forms a valley encircling the Black Hills. Interstate 90 follows this valley around the north end of the Black Hills from Spearfish, South Dakota, to Sundance. This is a good example of how different rock layers respond to weathering and erosion. Shales are typically very soft and easily eroded by streams, whereas sandstones and limestones are more resistant, and form bold outcrops.

The Bear Lodge Mountains, immediately north of Sundance, are composed of igneous rocks of Eocene age (53 to 55 million years old) that intruded and domed the older Paleozoic and Mesozoic strata. This type of mushroom-shaped igneous body is called a "laccolith." Black Buttes and Inyan Kara Mountain south of Sundance are also Eocene intrusive bodies. The town of Sundance lies at the foot of Sundance Mountain, known as the "Temple of the Sioux." Indians held a yearly ceremonial dance here to worship the sun. The "Sundance Kid," Butch Cassidy's outlaw partner, actually lived near here

Devils Tower and Missouri Buttes.

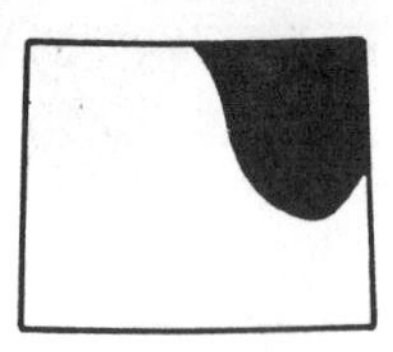

Devils Tower in the northeastern Powder River Basin.

in the 1800s. His real name was Harry Longabaugh and he spent 18 months in the local county jail for horse stealing.

DEVILS TOWER

Between Sundance and Moorcroft, Devils Tower can be seen to the north. However, for a worthwhile side-trip, take US 14 to Devils Tower National Monument. The Tower was called "Bad God's Tower" by the Indians, and was the first site to be declared a national monument in 1906. It rises 1,280 feet above the valley to an elevation of 5,117 feet above sea level. The Tower is distinctive due to the vertical fractures or "columnar joints" on its sides. These joints formed during cooling and contraction of the magma, much like the contraction of mudcracks in a drying mudpuddle. Indians believed the joints were caused by a great bear who scratched his claws on the side of the tower. The tower is believed by most geologists to be the

Close-up of columnar joints at Devil s Tower.

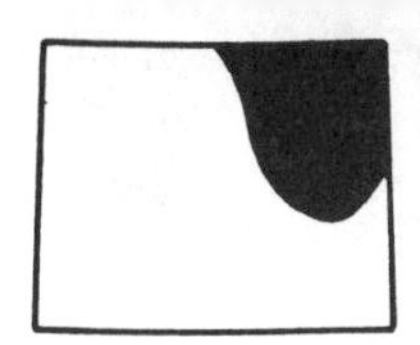

Devils Tower close-up of lithology.

eroded "throat" or neck of an ancient volcano. The Missouri Buttes, along the Little Missouri River north of Devils Tower, are also igneous bodies related to Devils Tower and formed about the same time.

Between Sundance and Moorcroft, Interstate 90 traverses the western flank of the Black Hills through west-dipping Jurassic and Cretaceous sedimentary strata. Some of the more noteworthy formations include the green-gray sandstone of the Jurassic Sundance formation, about 5 miles west of Sundance; rusty sandstones and variegated claystones of the lower Cretaceous Inyan Kara group on either side of Inyan Kara Creek; the black, soft Cretaceous shales belonging to several formations due south of Keyhole Reservoir; and the light-colored Fox Hills sandstone about 2 miles east of Moorcroft. All of these sandstones and shales were deposited in or along a shallow, marine seaway that covered the western interior of the U.S. throughout Cretaceous time. This seaway was so vast it connected the Gulf of Mexico with the Arctic Ocean! The sandstones were

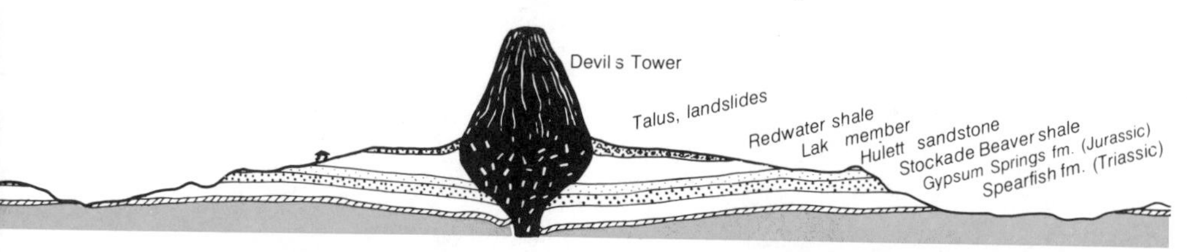

Cross section of stratigraphy in Devils Tower National Monument. The Tower is 1000 feet in diameter at the base, and rises 1267 feet above river level.

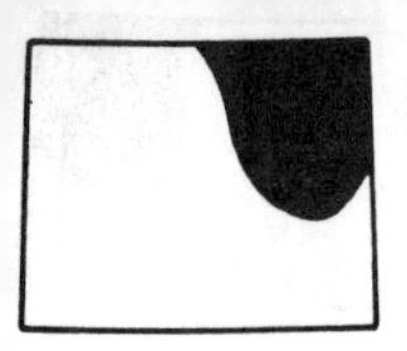

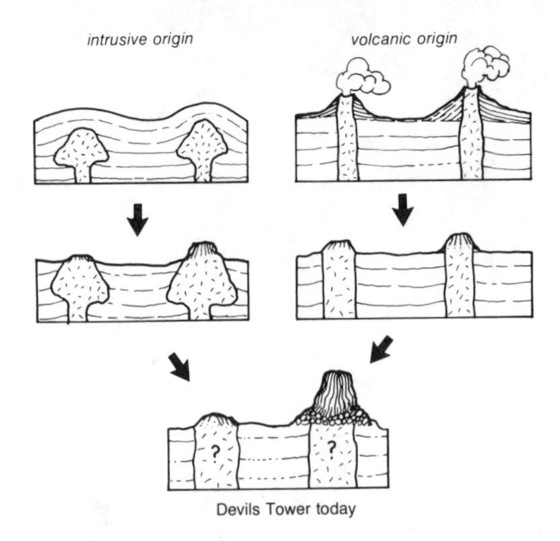

Two ideas for the formation of Devils Tower and Missouri Buttes. A: That they were originally intrusive bodies that did not completely penetrate the sedimentary section. B: That they are the eroded remnants of volcanic necks.

deposited as beach, delta and offshore bars of sand, while the black shales were deposited in quiet, deeper water further from shore.

Moorcroft is on the eastern edge of the Powder River Basin, where strata of early Cenozoic age lap onto the older formations that surrounded the Black Hills. The town stands on the upper Cretaceous Lance formation, a series of thick-bedded buff sandstones and greenish shales. The Lance formation consists of strata that filled in the Cretaceous marine seaway.

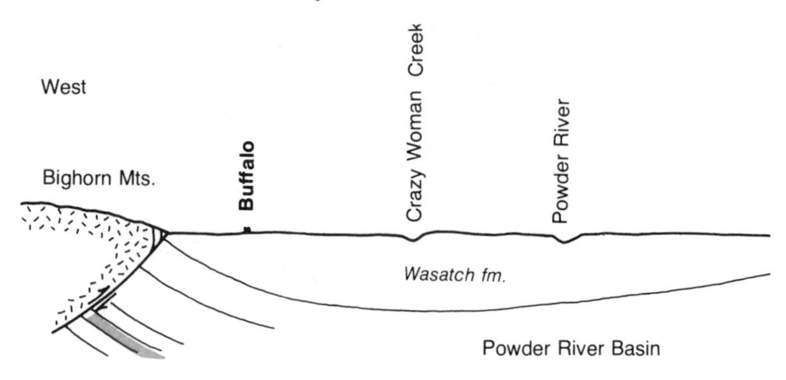

Geologic section along Interstate 90 from Wyoming/S. Dakota state line to the Bighorn Mountains, illustrating the stratigraphy and structure of the north end of the Powder River Basin.

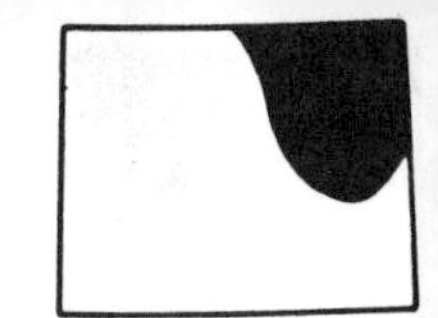

Resistant clinker beds of burned-out coal cap the hills.

Between Moorcroft and Gillette, Interstate 90 traverses basin-fill, non-marine sedimentary rocks belonging to the Paleocene Fort Union formation and the Eocene Wasatch formation. Numerous oil fields are scattered throughout this part of the basin. The Fort Union is the major coal producing formation in the eastern Powder River Basin. It crops out along most of the distance between Moorcroft and Gillette; watch for yellow sandstones interbedded with gray shales and coal beds. Gillette is on the Wasatch formation, drab sandstones and variegated claystones.

Gillette was named after Weston Gillette, a civil engineer who directed construction of the Burlington Railroad in 1891. Previous names for the town were "Donkey Town," after Donkey Creek which flows nearby, and "Rocky Pile," after Rocky Draw. The first airplane flown in Wyoming was at Gillette on July 4, 1911. Today, Gillette is a major energy center for the northern Powder River Basin, with numerous coal mines and oil fields in the vicinity.

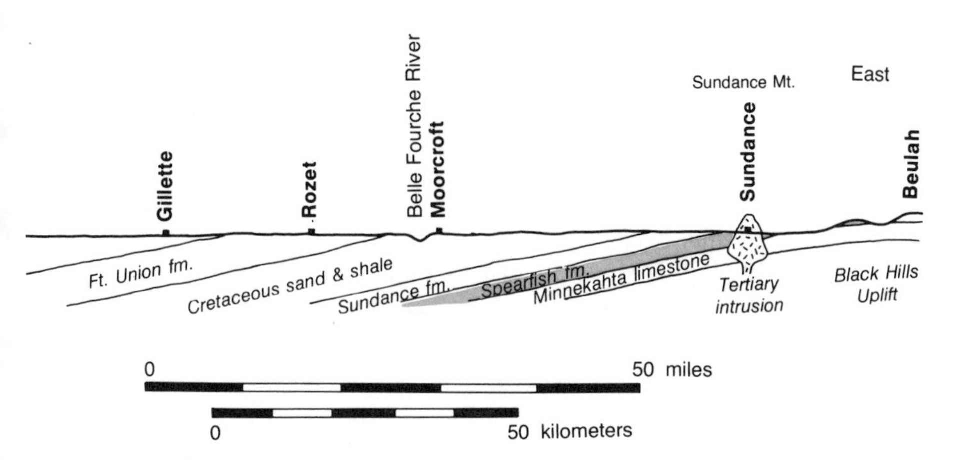

Interstate 90: Gillette—Buffalo
95 mi./152 km.

Interstate 90 crosses the vast expanse of the central Powder River Basin between Gillette and Buffalo. The floor of the basin is covered by the flat-lying Eocene Wasatch formation, noted for its drab sandstones, variegated claystones, and interbedded coals. It was deposited by streams around 50 million years ago, after retreat of the marine Cretaceous seaway and the Laramide orogeny. The rolling hills of Wasatch strata, covered with grass and sage, form a classic Wyoming basin scene.

The unusual "haystack" topography near Gillette is caused by resistant beds of "clinker," or burned coal and clay, that cap individual haystacks. Erosion dissected the land into isolated buttes and hills, each protected by a remnant of clinker at the top.

As you drive across the basin to the west, the Pumpkin Buttes will eventually be outlined on the horizon to the south, and the Bighorn Mountains will rise on the western horizon.

The north-flowing Powder River is crossed about halfway across the basin, where good outcrops of coal, siltstone and sandstone appear on the west bank of the river. They provide a detailed view of the sediments that filled Wyoming's basins during Cenozoic time.

Crazy Woman Creek is crossed about halfway between the Powder River and Buffalo. It was named for a crazy lady, but which one is not entirely clear. One story maintains that a young, white wife of a trader went crazy and lived in the mountains after her husband was killed by Indians. Another story says that an Indian woman went crazy after her village was attacked, and she lived her life in a squalid wikiup until her death. Whichever is true, the difficulty of early life in the Powder River country is immortalized by this creek.

The town of Buffalo stands at the base of the Bighorn Mountains at the junction of interstates 90 and 25. It was one of the first settlements in northern Wyoming; its main street supposedly follows an old buffalo trail. Such trails were used by the great herds for centuries, and many are still visible from airplanes.

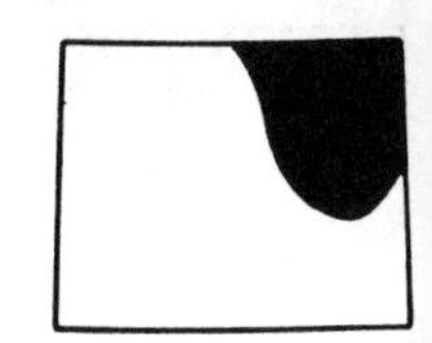

US 16: Newcastle—Moorcroft
44 mi./70 km.

Newcastle is on the southwest flank of the Black Hills uplift. Like Interstate 90 to the north, US 16 to Moorcroft traverses the western shoulder of the Black Hills through a succession of Mesozoic age sandstones and shales.

Newcastle was established in 1889 at the end of the Burlington Railroad. It was named after the English coal port of Newcastle-upon-Tyne because of coal deposits in the area. Today, Newcastle is an oil, lumber, and livestock center for the eastern Powder River Basin. Geologically, the town is on soft, black shales of the Cretaceous Belle Fourche formation.

US 16 follows the Belle Fourche shale for most of the distance between Newcastle and Moorcroft. Bentonite, a special type of clay used in industry, is mined near Upton. About ten miles north of Upton, the road leaves the Belle Fourche shale and cuts across uppermost Cretaceous strata toward Moorcroft, including the Fox Hills sandstone. It appears two miles southeast of Moorcroft, and consists of light-colored sandstone and gray, sandy shale full of marine fossils.

US 18/85: Lusk—Newcastle
81 mi./130 km.

Lusk is in the southeast corner of the Powder River Basin on an uplifted structure called the Hartville uplift. In a nutshell, the Hartville uplift is a structural arch that connects the Laramie Range with the southern Black Hills. Older rocks are intermittently exposed along the arch, but are covered around Lusk with a veneer of young, upper Oligocene and Miocene sediments called the Arikaree formation. The Arikaree consists of pale, soft sandstone and white claystone full of volcanic ash. These sediments were deposited by low-gradient streams that sluggishly meandered across a relatively flat Wyoming landscape 25 to 20 million years ago. The Arikaree formation extends north from Lusk to Hat Creek, an important stop on the historic Cheyenne-Deadwood Stage route.

For ten miles north of Hat Creek, US 85 traverses the White River formation deposited in the Oligocene, 35 to 31 million years ago. The White River, which underlies the Arikaree formation, is very distinctive and is composed of white to pale pink, blocky, tuffaceous claystone with lenses of conglomerate. The conglomerate was deposited in stream channels that cut through a floodplain of claystone. Abundant volcanic ash in the White River and Arikaree was probably blown from Nevada by prevailing westerly winds. Large volcanic eruptions of rhyolite ash occurred there at this time.

One of North America's first dinosaur discoveries was along Lance Creek, northwest of the junction of US 18/85 and Wyoming 270. Remains of Triceratops, a three-horned dinosaur, were found in the sandstones and shales of the upper Lance formation, which was deposited in swamps and rivers in latest Cretaceous time, just before the global extinction of dinosaurs.

From ten miles north of Hat Creek all the way to Newcastle, US 85 follows upper Cretaceous marine sedimentary rocks, especially the Pierre formation, a dark gray, marine shale containing numerous beds of bentonite clay. The Pierre was deposited about 75 million years ago at the bottom of a shallow marine basin that covered the western interior states and Alberta. Also, the upper Cretaceous Fox Hills sandstone crops out south of Mule Creek Junction, the junction of US 85 with US 18. The Fox Hills is light-colored sandstone interlayered with gray sandy shale.

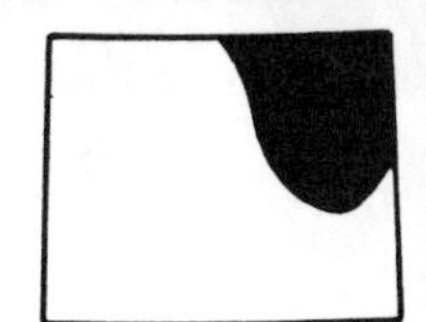

US 18
Lusk—Orin at Interstate 25
41 mi./66 km.

In the 1880s, Lusk was at the bustling crossroads of the Texas Trail and Cheyenne-Deadwood Stage route. Thousands of cattle were driven from Texas to Wyoming and Montana across the Powder River Basin during the last quarter of the 19th century. The Cheyenne-Deadwood Stage served the gold rush camps of the Black Hills, hauling passengers, mail and gold to and from the railhead in Cheyenne. First-class fare on the stage was $20, while third-class passengers paid just $10 — the catch was, they had to get off and push the stage over bad spots and up hills! Calamity Jane and Wild Bill Hickok were frequent passengers on the stage.

The road between Lusk and Orin (at Interstate 25) traverses young sediments deposited on the northwest flank of the Hartville Uplift. The entire route crosses the upper Oligocene and Miocene Arikaree formation, between 25 and 20 million years old. It consists of light-colored, soft, porous sandstone and white, tuffaceous claystone. Tuffaceous means that the rock contains volcanic ash, in this case blown in during the Miocene from distant volcanoes to the west.

WESTERN POWDER RIVER BASIN

For continuity, the section of Interstate 25 between Douglas and Casper is presented in the chapter on southeastern Wyoming. This way, travelers on Interstate 25 between Casper and Cheyenne have a complete, uninterrupted discussion of the geology.

Interstates 25 & 90

Casper—Buffalo (Interstate 25) 112 mi./179 km.

Buffalo—Sheridan (Interstate 90) 35 mi./56 km.

This section is divided into two segments based on the age of bedrock formations. The first segment, between Casper and the Midwest-Edgerton junction, crosses upper Cretaceous strata on the

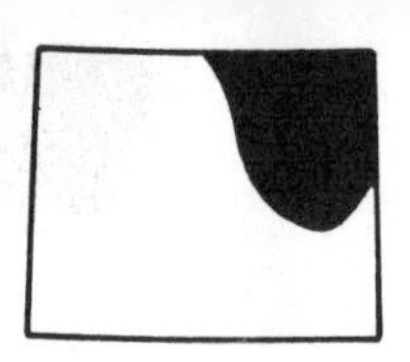

Teapot Rock on Highway 259, 10 miles south of Midwest and Edgerton.

southwestern shoulder of the Powder River Basin. The second traverses younger Cenozoic strata in the central trough of the Powder River Basin.

Interstate 25 leaves Casper on the dull gray shales of the Cretaceous Cody formation, and reenters Cody shales in the middle of Salt Creek oil field near Midwest. In between, Interstate 25 cuts through strata belonging to the Mesaverde sandstone, which contains the Teapot sandstone which forms isolated buttes and knobs of white to tan, massively-bedded sandstone, Lewis shale, Fox Hills sandstone, and Lance sandstone. All were deposited in late Cretaceous time, about 80-65 million years ago. The Mesaverde, Lewis and Fox Hills formations are marine sediments that were deposited within or along the shore of a shallow Cretaceous seaway. The Lance formation is a non-marine, river-deposited unit that eventually filled in this seaway in latest Cretaceous time. These strata are locally covered by geologically recent sand dune deposits that blew east from the Wind River Basin over the Casper arch.

For a geological side-trip, take the Wyoming 259/387 loop through Midwest and Edgerton, to the heart of two of Wyoming's largest and most controversial oil fields — Teapot Dome and Salt Creek.

Wyoming 259 cuts through outcrops of the Teapot sandstone north of the Interstate 25 junction. This formation, named for a prominent outcrop just east of the highway, resembled a teapot — unfortunately, a lightning storm damaged the handle of the teapot in 1962, so it is not so obvious today.

The Teapot Dome field, just south of Salt Creek, was the stage for a major political scandal of near-Watergate proportions. Teapot Dome, discovered in 1922, was established as a U.S. Naval Petroleum Reserve, since a peace time reserve of oil was needed for ships. During the administration of President Warren G. Harding, Albert Fall of New Mexico was appointed Secretary of the Interior. While in office,

Fall leased government oil land at Teapot Dome and in California to his associates, Harry Sinclair and Edward Dohney, after accepting a $100,000 bribe. When this was discovered, a political storm swept through the Harding administration. Fall became the scapegoat for the entire administration, resigned as Secretary of the Interior and was sent to prison in 1931. He died in poverty in 1944 in El Paso, Texas.

The Salt Creek field is one of the world's classic oil fields. The oil accumulated along the crest of a large anticlinal fold that is expressed at the surface by gently dipping beds of buff-colored, Cretaceous Shannon sandstone that have eroded back to form prominent cliffs up to 100 feet high. Gray shales of the older Cody formation are exposed in the center of the fold. The first oil well in the area was drilled in the fall of 1889 by P.M. Shannon of the Pennsylvania Oil and Gas Company, north of the present field. The oil was sent to Pittsburgh for analysis where it was initially pronounced "unnatural" because such light crude had never been found before! More wells were dug and for years the oil sold for $10.00 per barrel and was used by the Union Pacific Railroad in its unrefined state as a lubricant. Transportation of the oil was a tremendous problem — it was hauled 50 miles to Casper by teams of 12 to 15 horses, each pulling a wagon with the equivalent of 45 barrels. It wasn't until 20 years later that people realized that this wagon road passed right over the crest of the Salt Creek dome where the real wealth of oil lay buried! In 1908, the first well of the Salt Creek field was completed — it was a tremendous gusher and ushered in Wyoming's first, major oil boom. By 1985, the Salt Creek field alone, neglecting adjacent satellite fields, had produced over 610,000,000 barrels of oil and over 711,000,000 MCF of gas! Wyoming 259 passes through the heart of the field, past modern-day "pump-jacks" and former drilling derricks.

The Rimrock (Shannon sandstone) surrounds Salt Creek anticline and oil field near Midwest and Edgerton, north of Casper. Best seen from Highway 259.

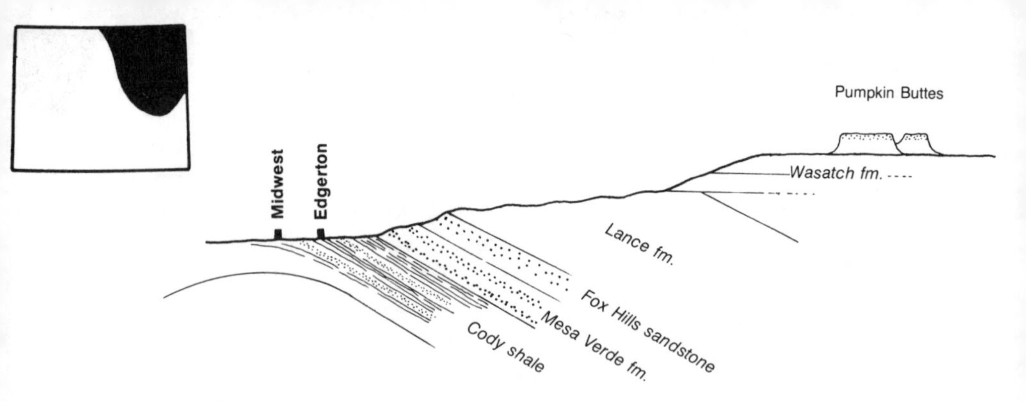

Cross section of stratigraphy from Midwest to Pumpkin Buttes along Highway 387.

North of the Midwest-Edgerton junction, Interstate 25 follows the eastern flank of the Bighorn Mountains. The vast Powder River Basin stretches to the eastern horizon, with bluffs and hills looking like a choppy sea. The interstate follows the upper Cretaceous Cody shale from the Midwest-Edgerton junction to Kaycee. At Tisdale Mountain Road, the Tisdale dome is outlined by resistant, dark-brown sandstone members of the Cretaceous Frontier formation on the west side of the highway. The flanks of the dome have been dissected by streams to form several small, distinctive canyons. This anticlinal structure has produced more than 6,000,000 barrels of oil, and some gas, from upper Paleozoic and lower Mesozoic strata. To the east, drab shales of the Cody formation form hilly topography. The South Fork of the Powder River dissects the north end of Tisdale dome at Powder River Road.

At Kaycee, the Cretaceous Frontier sandstone dips eastward into the basin to form classic "hogbacks" or ridges. Kaycee was named after the K-C brand of a local ranch, dating back to the 1880s. Immediately north of Kaycee, the interstate passes by outcrops of yellow, massive sandstone of the Mesaverde formation. Farther north, Interstate 25 encounters the Paleocene Fort Union formation and follows it for the next 20 miles or so. It is recognized by its drab-tan shales, sandstones and interbedded coals which form rolling hills. To the west, the "Horn" is a distinctive mountain formed by Precambrian rocks that project south from the main mass of the Bighorns. The Horn is bound by a thrust fault on its west side, and was uplifted and shoved westward.

Interstate 25 passes through the Cellars Ranch oil field just north of Kaycee. This field has produced over 5,000,000 barrels of oil from the Pennsylvanian Tensleep sandstone since its discovery in 1960. The Tensleep sandstone is exposed along the flank of the Bighorn

Range to the west, dips steeply into the subsurface of the Powder River Basin, and is several thousand feet below ground level at Cellars Ranch field.

The area between Kaycee and Buffalo was supposedly the site of the mysterious "Lost Cabin Mine." There are many versions of the tale, but basically before the Civil War a party of Swedes supposedly found a fantastic gold deposit somewhere in the Bighorn Mountains and built a cabin at the site. They were attacked by Indians, and all but two were killed. These two made it to Fort Laramie with $7,000 in gold, reportedly mined in just 3 days. The Swedes were never heard from again and the cabin has never been found.

At Middle Fork Road the interstate climbs onto a high surface formed by the Eocene Wasatch formation. The Wasatch is exposed along Interstate 90 between Buffalo and Sheridan. Watch for its typically variegated claystone, drab sandstones, and interbedded coals in the lower part. The Wasatch formation is the dominant coal-bearing unit in the western Powder River Basin, and contains some of the world's thickest coal seams. The Lake De Smet coal seam stretches north of Buffalo for over 15 miles — it is 1 to 2 miles wide and 250 feet thick in places! Thinner coal beds are exposed in roadcuts adjacent to Interstate 90 from the north end of Lake De Smet to Sheridan. The swampy environment that produced the coal was a narrow trough or "moat" next to the rising Bighorn Mountains during the Laramide orogeny. As the mountains rose and the basin sagged, these incredibly thick coal beds were deposited parallel to the eastern flank of the mountains. Lake De Smet is named for Father Pierre Jean De Smet, who saw the lake in 1840 with a party of fur trappers.

In addition to coal, beds of bright red "clinker" appear along this section of Interstate 90. Clinker formed when exposed coals beds were ignited, perhaps by lightning, and slowly burned, baking the surrounding, interbedded shales and sandstones to form a rough, blood-red slag. Clinker is locally mined for road gravel.

Also between Buffalo and Sheridan, the central mass of the Bighorn Mountains looms to the west. This is the highest part of the range, capped by the summit of Cloud Peak at 13,165 feet. This section of the Bighorns, called the Piney Creek block, has been thrust faulted eastward over the margin of the Powder River Basin. There is over 30,000 feet of structural relief here between Precambrian rocks exposed in the mountains and the same rocks buried beneath sedi-

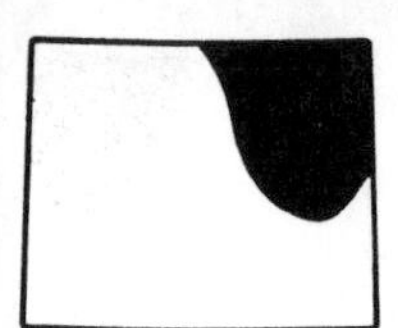

ments of the adjacent Powder River Basin. The Laramide mountain building episode in early Cenozoic time not only compressed and squeezed the crust, it also caused tremendous vertical displacement between the mountains and basins.

Sheridan is near the contact between the Paleocene Fort Union and Eocene Wasatch formations. The Fort Union extends north beyond the state line and west to the foot of the mountains, while the Wasatch extends eastward over the northern Powder River Basin. The Bighorn Coal Mine, north of Sheridan, strips coal from the upper part of the Fort Union formation. Sheridan was named after a Civil War general, and is the lowest town in the state at 3,745 feet. There is 9,420 feet of topographic relief between downtown Sheridan and the summit of Cloud Peak 30 miles to the southwest!

CROSSING THE BIGHORN MOUNTAINS

Two routes cross the backbone of the Bighorn Mountains. The northern route connects Sheridan with Lovell and Greybull, while the southern route connects Buffalo with Worland. Both are wide, well traveled highways that offer spectacular views of the mountains and adjacent basins. Both roads rise to over 10,000 feet, so if you come from sea level, be prepared for thin air!

US 14 and Alternate 14: Northern Crossing Ranchester—Burgess Jct.

The northern crossing of the Bighorns begins at Ranchester on the eastern side, about 10 miles north of Sheridan, or at Lovell or Greybull on the western side in the Bighorn Basin.

Between Ranchester and Dayton the road follows the Tongue River as it meanders through the Paleocene Fort Union formation. West of Dayton, the road gradually climbs through Mesozoic formations of shale and sandstone. In particular, the distinctive red shales of the Triassic Chugwater formation provide a bright splash of color at the base of the range. The main, steep face of the range is formed by resistant sandstone and limestone strata of upper Paleozoic age, such

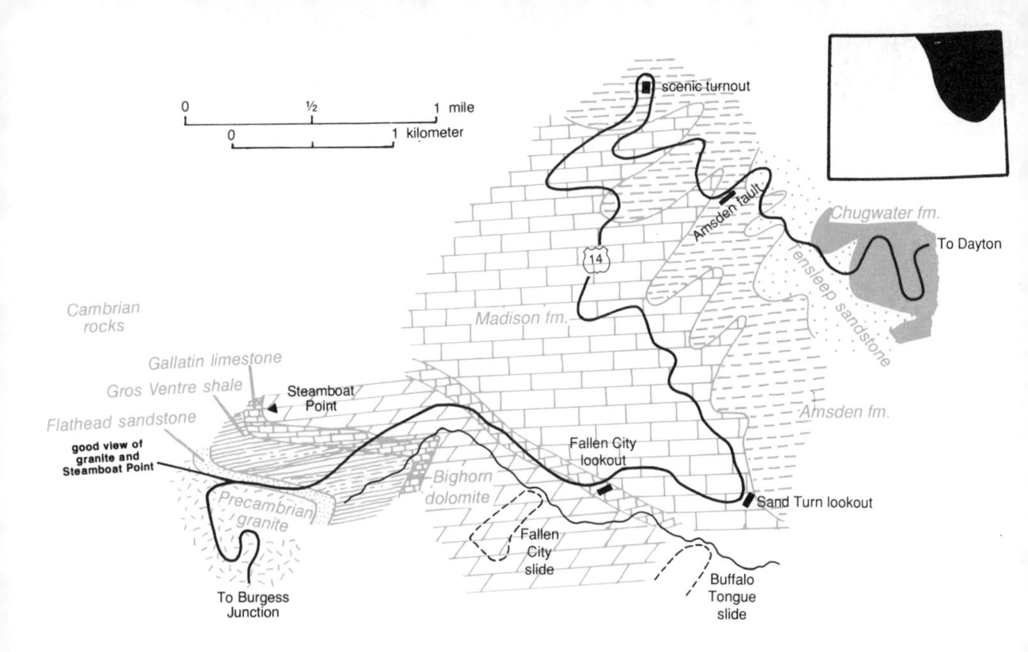

Geologic map of the east flank of Bighorn Mountains along US 14.

as the Pennsylvanian Tensleep sandstones and gray limestones of the Mississippian Madison formation.

The "Fallen City" is a spectacular rock slide in the Madison limestone, where slabs of limestone have been pulled downhill by gravity along steeply inclined bedding surfaces. Near the top, Steamboat Rock is a prominent outcropping of the Ordovician Bighorn dolomite. The Bighorn dolomite is a distinctive formation recognized by its massively bedded, white to light gray dolomite. As you drive across the top of the range, you will see several erosional remnants of Bighorn dolomite and younger strata, usually forming flat-topped buttes; these outcrops are all that remains of the Paleozoic and Mesozoic marine sedimentary rocks that have been stripped off the top of the range by erosion. These rocks were deposited long before the

Cross section view of east flank of Bighorn Mountains from highway US-14.

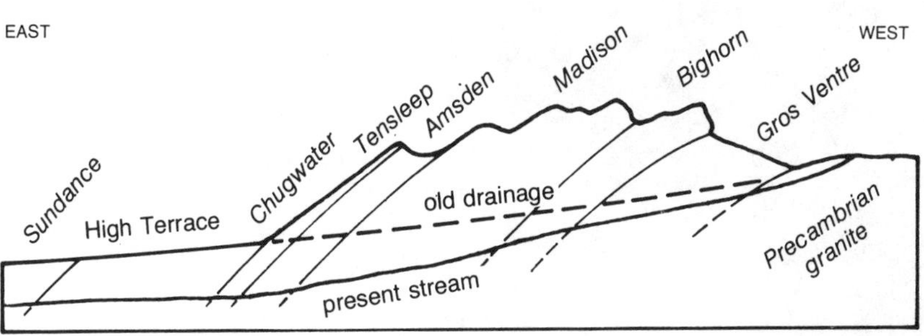

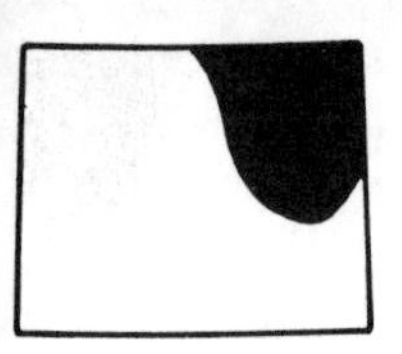

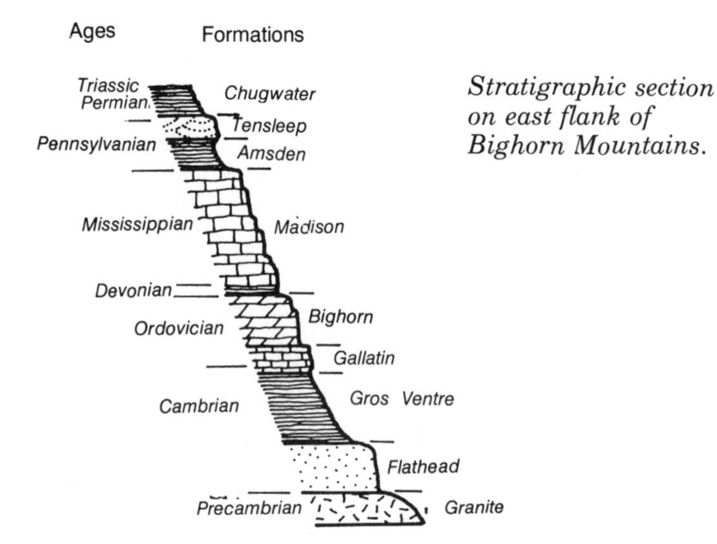

Stratigraphic section on east flank of Bighorn Mountains.

A

Small fault cutting the Mississippian Amsden formation along Highway 14 west of Dayton, east flank of Bighorn Range.

B

"Fallen City" rock slide in Madison limestone along Highway 14 west of Dayton.

Rocky Mountains existed, at the bottom of shallow marine seas that covered western North America from 600 million to 65 million years ago.

Between Steamboat Rock and Burgess Junction, the road traverses ancient Precambrian igneous rocks in the core of the range. These rocks are approximately 2.9 billion years old! Their composition is in between that of a granite and a basalt, a special type of igneous rock geologists call a "quartz diorite."

Alternate 14: Burgess Junction—Lovell

58 mi./93 km.

At Burgess Junction, US 14 splits — the south arm goes to Greybull and the north arm heads to Lovell. Both are scenic, excellent roads, as are most Wyoming roads! West of Burgess Junction, Alternate 14 follows a broad valley cut by the North Fork of the Tongue River. It was cut deeply into the range by a process called "headward erosion." This simply means that the headwaters of the river are gradually eroding back farther into the range. The Cambrian Flathead sandstone and Gros Ventre shale crop out next to the road; the higher cliffs to the north and south are the white-gray Ordovician Bighorn dolomite. These Cambrian strata are around 530 million years old and contain fossils of the first, hard-shelled marine organisms found

Paleozoic sedimentary rocks dipping eastward on the east flank of the Bighorn Mountains, as seen from Highway 14 west of Dayton.

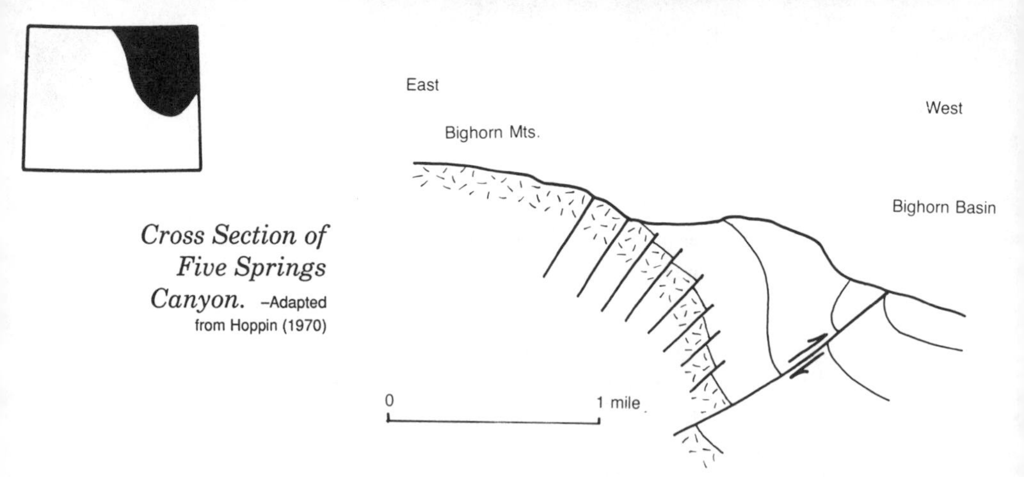

Cross Section of Five Springs Canyon. –Adapted from Hoppin (1970)

in the stratigraphic record; before this time, life on this planet did not have "hard-parts" or shells that could be fossilized. As you drive west, notice that the Tongue River gradually cuts deeper into Cambrian strata until, about 14 miles west of Burgess Junction, it finally hits Precambrian "basement" rocks, where the road turns sharply north.

Alternate 14 follows the western rim of the Bighorn Range on hard, resistant Precambrian rocks before it descends the steep west flank of the range. The arid floor of the Bighorn Basin is panoramically spread out to the west; on a clear day, you can see the Absaroka Range west of Cody, over 60 miles away. Stop your car at one of the many turnoffs and take in the view — everything spread before you has taken the Earth four and a half billion years to create!

The Indian Medicine Wheel is a fascinating place to stop along the western rim. This stone pattern was constructed by ancient Indians that pre-dated the tribes of the 1800s. No one knows exactly why it was built or what it means, but it may be a calendar or religious site.

Geologically, the west flank of the Bighorn Range is almost a mirror image of the east flank with respect to the rock strata. The

West flank at Bighorn Mountain and the eastern Bighorn Basin as seen along Highway 14-Alt. between Burgess Junction and Lovell.

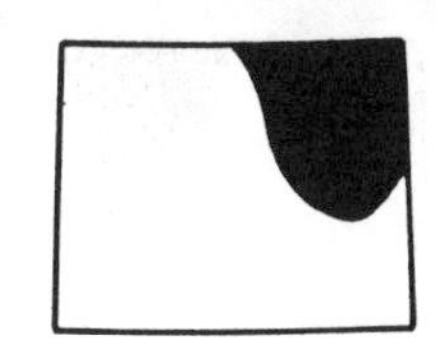

Overturned Paleozoic and Mesozoic strata in hanging wall at Five Springs thrust, west flank of Bighorn Mountains along Highway 14-Alt.

road cuts down through resistant cliffs of Bighorn dolomite, Madison limestone, and Tensleep sandstone to the softer Mesozoic strata at the base of the range. Here, the Five Springs thrust fault shoved the range westward over the Bighorn Basin, placing older Triassic and Jurassic rocks over black Cretaceous shales at the surface where the road grade markedly flattens. From the base of the range to Bighorn Lake, Alternate 14 traverses soft Cretaceous black shales that form a broad valley for the Bighorn River. Between the lake and Lovell, the road passes by the nose of Little Sheep Mountain anticline, which forms a prominent hill south of the highway. Like most anticlines in the Bighorn Basin, the bright red Triassic Chugwater formation forms a distinctive red band around the perimeter of the structure.

Wyoming 37 leads to the Bighorn Canyon National Recreation Area, where the Bighorn River slices deeply into rock layers along the north end of the Bighorn Mountains. Over 500 million years of Earth history is exposed along the vertical rock walls that tower above the river. Yellowtail Dam, in Montana, creates a lake that extends 71 miles from northern Wyoming into Montana.

US 14: Burgess Junction—Greybull
48 mi./77 km.

US 14 follows the beautiful Shell Canyon from Burgess Junction to Greybull. This route may be divided into two segments, a north-south segment south of Burgess Junction, and an east-west segment through Shell Canyon to Greybull.

The north-south segment traverses the high, flat-topped "sub-summit surface" of the Bighorn Range. Precambrian igneous rocks

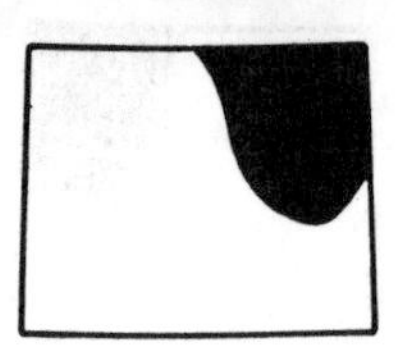

Cross section sketch of Shell Canyon.

West
East
Block of Paleozoic sandstone & limestone
thickened Cambrian shale
Precambrian
0
1 mile

crop out east of the road; the Cambrian Flathead sandstone and Gros Ventre shale form the tree-covered ridge to the west. Higher peaks are located to the southeast along the range crest, but this plateau-like surface is typical of most of the range. Adjacent basins were filled with sediment to the level of this surface until about 5 million years ago, when regional uplift caused rivers to downcut their channels and exhume the old mountains and basins. During the recent Ice Ages, glaciers accumulated along the range crest and spread over this surface, scouring and eroding it down.

Shell Canyon has been cut into the top of the range by the headward erosion of Shell Creek. The creek sliced a deep, narrow gorge into Precambrian rocks at the rest area, about half way down the canyon. At the rest area, you can see outcrops of the Cambrian Flathead sandstone, about 550 million years old, resting on 2.9 billion year old Precambrian rocks. These Precambrian rocks are some of the oldest rocks on earth, while the Cambrian sandstones contain the first evidence of fossilized hard shells! These mountains are made of old stuff! Paleozoic strata dip rather gently to the west along the high cliffs adjacent to the canyon, except near the canyon mouth where they sharply bend to a steep angle and plunge into the Bighorn Basin. The canyon mouth is formed by resistant cliffs of Ordovician Bighorn dolomite, Mississippian Madison limestone, and Pennsylvanian Tensleep sandstone, all of which dip steeply into the Bighorn Basin.

West of the mountain front toward Greybull, US 14 traverses the

Cross-bedded Tensleep sandstone along Highway 16, west flank of the Bighorn Mountains.

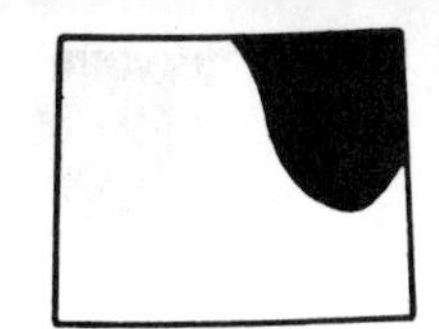

Panoramic view of Precambrian rocks along the crest of the southern Bighorn Mountains between Buffalo and Tensleep along Highway 16.

eastern shoulder of the Bighorn Basin on soft Mesozoic shales and sandstones. The bright red shales between the canyon mouth and the town of Shell belong to (guess what?) the Triassic Chugwater formation, which also forms a distinctive red outcrop on the east side of the range. The town of Shell was probably named for the numerous shell fossils found at the mouth of Shell Creek where it emerges from the canyon. From Shell to Greybull the road crosses black, marine shales and sandstones of Cretaceous age. Sheep Mountain anticline forms a prominent ridge just north of Greybull.

US 16: Southern Crossing— Buffalo—Worland
90 mi./144 km.

The southern route over the Bighorn Mountains is just as scenic as the northern route. From Buffalo to the eastern mountain front, the highway traverses Eocene rocks of the Wasatch, at the base of the range. The Kingsburg conglomerate member of the Wasatch, at the base of the range, is composed of cobbles of Paleozoic rocks that were deposited as gravels on alluvial fans along the front of the rising Bighorn Range around 55 million years ago. About 6 miles west of Buffalo, the road crosses the Piney Creek thrust fault, which shoved this portion of the central Bighorns over the western margin of the Powder River Basin. The road climbs through a narrow section of steeply inclined Paleozoic limestone and dolomite beds in the hanging wall of the Piney Creek thrust, then cuts into much older Precambrian "basement" rocks. The Precambrian is composed of extremely ancient metamorphic gneisses (pronounced "nices") that are over three billion years old. The Earth is 4.7 billion years old, so these

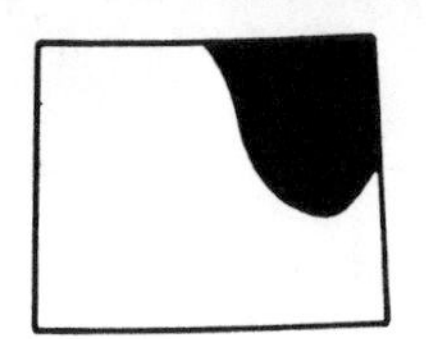

Folded metamorphic gneiss along Highway 16 between Buffalo and Tensleep.

rocks are about two-thirds of the age of the planet. These metamorphic rocks formed as extreme temperature and pressure transformed pre-existing rocks, probably during an ancient mountain-building episode, that converted them to metamorphic gneiss. It is likely that the pre-existing rocks were sedimentary layers of sandstone and shale that were deposited around 3.5 billion years ago. To think like a geologist, you must think in terms of millions and billions of years! Mother Nature has done great things with this planet, but she tends to take her time!

US 16 follows Precambrian rocks across the crest of the range to the west side, where it again cuts through Paleozoic sedimentary layers along Tensleep Canyon. The prominent cliff-forming units in Tensleep Canyon, from Meadowlark Lake west, are the white Ordovician Bighorn dolomite, gray limestones of the Mississippian Madison formation, and the yellow sandstone of the Pennsylvanian Tensleep formation. The town of Tensleep is on soft, red shales of the Permian Goose Egg formation. Tensleep was named by Indians because it was "ten sleeps" or nights between Fort Laramie and Yellowstone.

Between Tensleep and Worland, US 16 traverses the southeastern corner of the Bighorn Basin. The eastern half of this section of road is over upper Cretaceous marine shales and sandstones, whereas the western half is over the non-marine sediments of the Fort Union and Wasatch formations.

Paleozoic sedimentary section above Tensleep, Highway 16.

Cross section of Pumpkin Buttes–large mesas seen for miles in the Powder River Basin.

Wyoming 387/59:
Midwest—Wright—Gillette

Wyoming 387 cuts through uppermost Cretaceous sandstones and shales between Midwest and the junction with Wyoming 192, about 12 miles northeast of Midwest. The relatively high, hilly topography in this area is caused by resistant sandstone units within these upper Cretaceous formations, such as the Shannon, Sussex, and Teapot sandstones. At the junction with Wyoming 387 and 192, the non-marine Paleocene Fort Union formation is exposed in road cuts. This unit is composed of gray shales, concretionary sandstone, and thin beds of coal, which were deposited as floodplain and river sediments around 65 million years ago. East of here, Wyoming 387 traverses the non-marine Eocene Wasatch formation across the rolling, grass-covered floor of the basin. The Wasatch is distinguished throughout Wyoming for its red, variegated claystones and interbedded sandstones. It is the most important coal-bearing horizon in the western Powder River Basin, and also contains extensive uranium deposits in the southern basin. Wyoming 387-59 traverses Wasatch outcrops all the way to Gillette.

The Pumpkin Buttes are three, orange-colored, flat-topped buttes that rise to the northeast across the basin. They are erosional sandstone remnants of the Oligocene White River formation that once blanketed the area. The buttes were used as Indian lookouts and landmarks along the Bozeman Trail. Uranium was discovered near the buttes in 1951, leading to a flurry of exploration and later development.

The Bozeman Trail passes through this part of the Powder River Basin. The trail was blazed by John M. Bozeman from 1863 to 1865 as

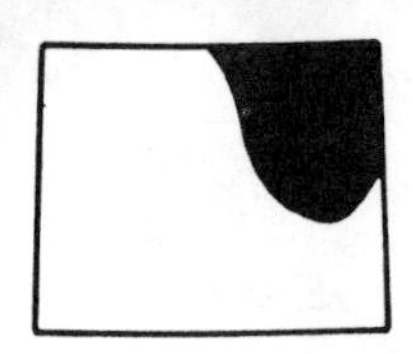

the shortest route to the gold fields of Virginia City, Montana. The trail pierced the heart of Sioux Indian hunting grounds and led to Red Cloud's war, one of the most successful Indian campaigns. The trail was abandoned for several years until the Sioux were suppressed. After 1877, the Bozeman Trail served as an important cattle route from Texas to Montana.

Between Wright and Gillette, Wyoming 59 passes numerous oil fields and leads to the heart of the Powder River Basin's coal mining district.

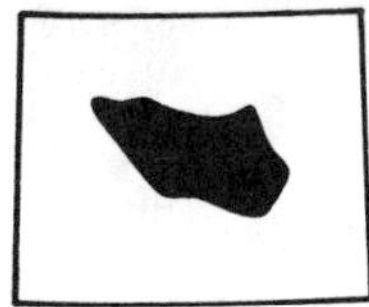

Central Wyoming — The Wind River Basin

INTRODUCTION

The Wind River Basin is the central heart of Wyoming. Sagebrush, wind, purple mountains and fast-moving rivers combine into a classic Wyoming landscape. The basin is a large, rhomb-shaped structural depression that covers all of Fremont County and the western part of Natrona County. Elevations average about 5,500 feet but range from 4,600 feet where the Wind River enters Wind River Canyon, to 13,785 feet on Gannett Peak in the Wind River Mountains, the highest peak in Wyoming. This area extends from the barren flats near Shoshoni to the dense forests and snow-capped peaks of the Wind River Range. Temperatures range from well below zero in winter to over 100° F in summer. Precipitation varies widely as well — Shoshoni averages a mere 5 inches per year, while the high peaks of the "Windies" receive more snow than melts in summer, forming active glaciers.

The basin's namesake is the Wind River, which originates on the eastern side of the Continental Divide at the north end of the Wind River Mountains. The Wind River swiftly flows east across the basin to Riverton, then sharply turns north into the Wind River Canyon and Bighorn Basin where it becomes the Bighorn River. The Wind River was so-named by Indians for the prevailing strong winds that blow down-stream between Absaroka and Wind River mountains. This phenomenon is quite common in the Rockies where cool air accumulates along

Cross section of Wind River Basin.

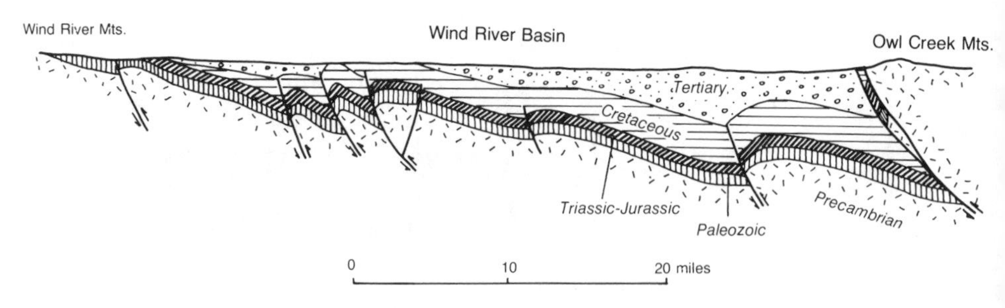

the high Continental Divide and flows down topographic avenues to create "down-canyon" winds.

Other major drainages in the basin include the Popo Agie (pronounced pō-pō-zha) and Sweetwater rivers. Popo Agie is a Crow Indian name for "head water," also called Wan-ze-Gara by Shoshone Indians. The Popo Agie River was the site of three fur trappers' rendezvous, one near Lander in 1829 and two at the confluence with the Wind River in 1830 and 1838; these were the only rendezvous held in the Wind River Basin. The Green River Basin was more popular. The Sweetwater River was probably named by General William H. Ashley in 1823, because of the purity and sweet taste of the water compared to the bitterly alkaline water of streams elsewhere; French speaking trappers called it Eau Sucree, or sugar water. The name may also be based on a story about a mule packed with a load of sugar that fell into the river, although Ashley was probably the real source of the name.

Structural uplifts frame the Wind River Basin. The massive Precambrian block of the Wind River Mountains forms the western side of the basin; the Washakie and Owl Creek mountains on the north separate it from the Bighorn Basin; the low Casper arch on the east divides it from the Powder River Basin; and the Granite Mountains define the south margin of the basin. The deepest part of the basin is along its northern and eastern margin, where adjacent uplifts have been thrust-faulted over the basin during the Laramide orogeny. As a result, these areas have the thickest accumulation of sedimentary rock, up to 20,000 feet.

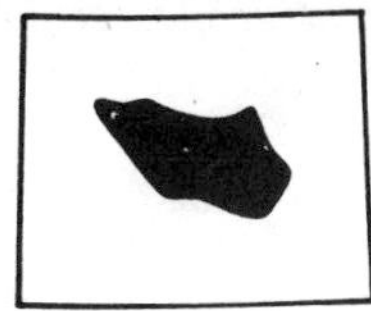

Most of the surface of the Wind River Basin, including the area between Shoshoni and Riverton, is floored by the Eocene Wind River formation. It is about 49 million years old, and contains brightly colored, variegated claystone and sandstone with some interbedded lenses of conglomerate. Much of the sediment was derived and reworked from volcanic source rocks in the Absaroka volcanic field, along the eastern side of Yellowstone National Park. It is much older than the recent activity of the Yellowstone volcano. Absaroka volcanism immediately followed the Laramide orogeny, so these volcanic strata generally lie flat on top of the folded and faulted rocks beneath.

SOUTHERN WIND RIVER BASIN

Granite Mountains

The Granite Mountains, also called the Sweetwater Hills, on the divide between the Wind River Basin and the Great Divide Basin, are a uniquely preserved landscape from Wyoming's geologic past. They were uplifted and thrust to the south during Paleocene time, when they became a high mountain range 90 miles long by 30 miles wide. They then resembled today's Wind River Mountains. Tertiary sediments, eroded from adjacent mountains, eventually blanketed most of Wyoming, filling in the Laramide basins and partially covering the mountains. In the Granite Mountains, the Miocene Splitrock formation eventually lapped over the central part of the range and covered the granite. Then in late Tertiary time the roof of the Granite Mountains collapsed along a series of normal faults to form a large trough. Regional uplift and erosion in the past two to three million years stripped much of the Tertiary sediment off the Rockies, but a two million year-old landscape remains almost completely preserved in the graben of the Granite Mountains. A trip across the Granite Mountains is like visiting Wyoming millions of years ago!

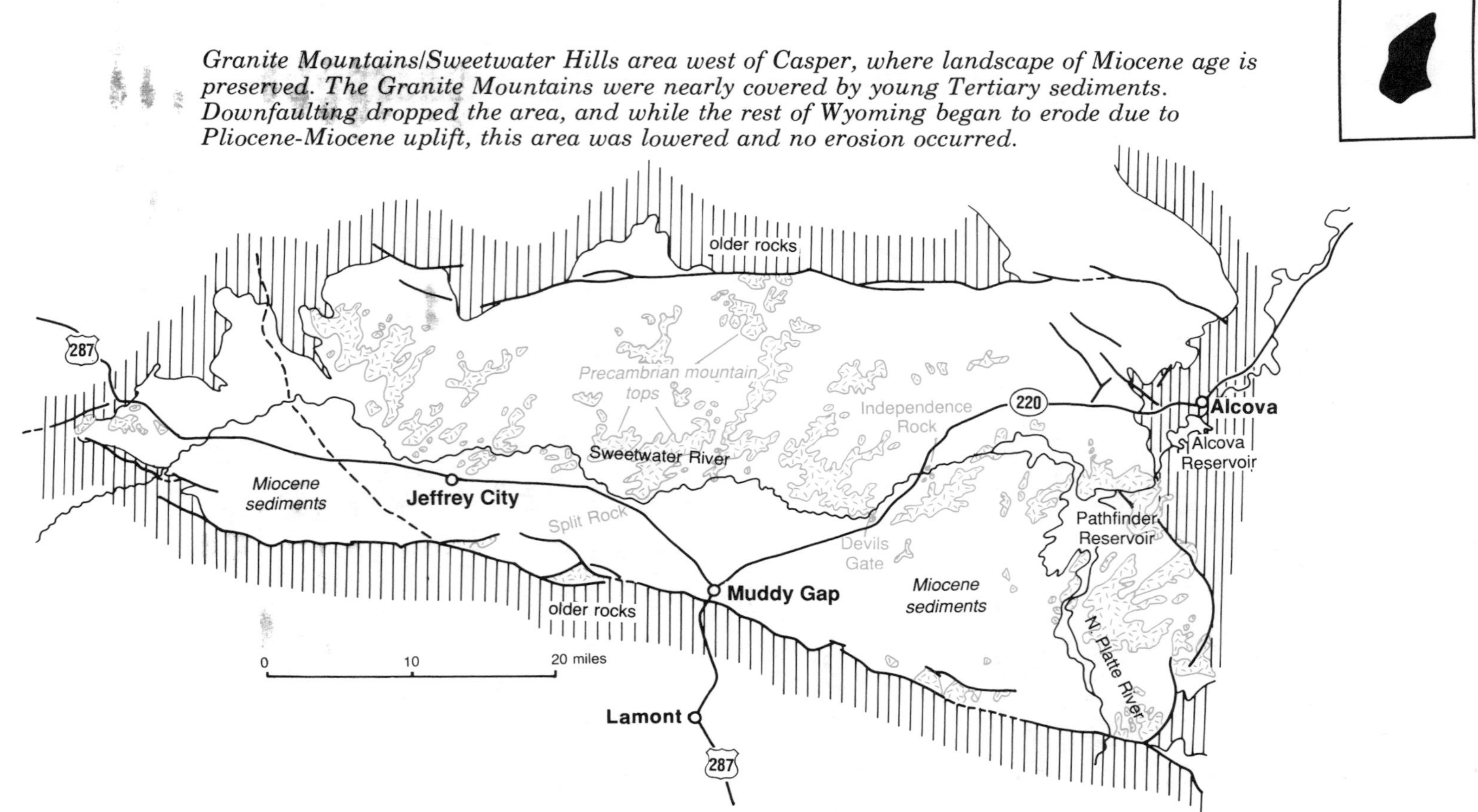

Granite Mountains/Sweetwater Hills area west of Casper, where landscape of Miocene age is preserved. The Granite Mountains were nearly covered by young Tertiary sediments. Downfaulting dropped the area, and while the rest of Wyoming began to erode due to Pliocene-Miocene uplift, this area was lowered and no erosion occurred.

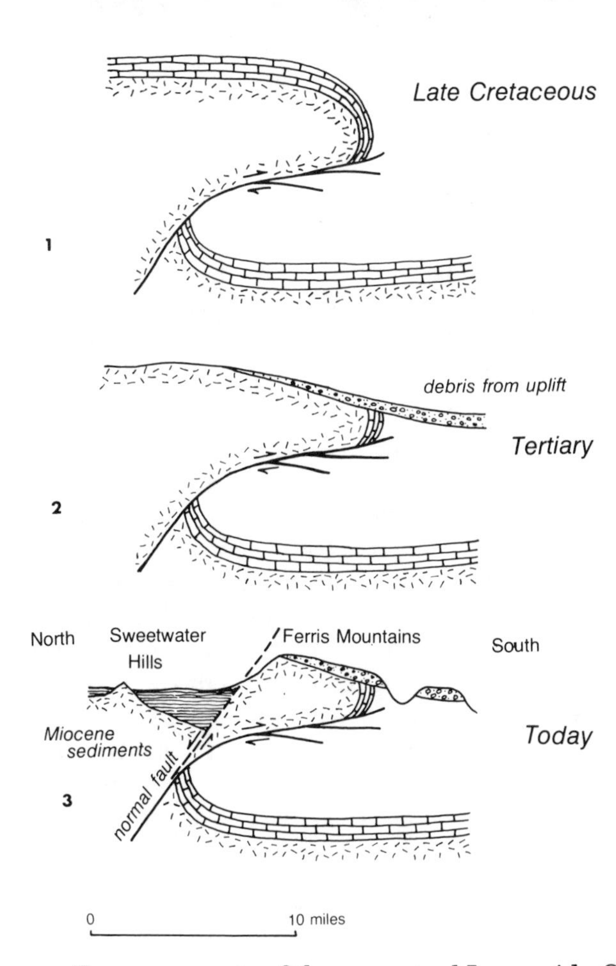

Development of Ferris Mountains and adjacent Sweetwater Hills through time. 1) Mountain core is thrust upward. 2) Erosion planes mountains, and debris is shed. 3) Collapse along normal fault and fill by Miocene sediments. –Adapted from Sales (1983, p. 83)

Two remnants of the ancestral Laramide Granite Mountains are left. On the south, Crooks Mountain, Green Mountain and the Ferris Mountains form a ridge of Precambrian granite and Paleozoic rocks that were thrust southward during the Laramide orogeny. On the north, the Rattlesnake Hills were part of the northern shoulder of this large mountain block. In between, the pink hills of granite that the Sweetwater River flows through today, are but a glimpse of the former crest of the uplift. The Sweetwater River was stopped in its easterly course through the central collapsed graben, so it could not flow north to join the Wind River.

The Granite Mountains are named for the extensive outcrops of red to pink granite. This rock contains pink orthoclase feldspar, quartz, and muscovite mica that cooled from magma to form an intrusive igneous rock around 1.8 billion years ago. Although granite is the predominant Precambrian rock in this

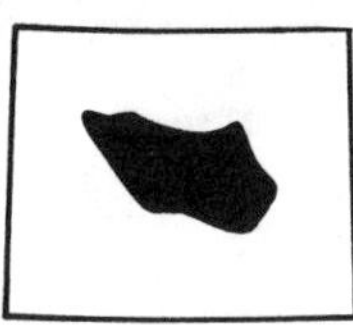

area, there are also exposures of older black and green schist, slate, and quartzite. These rocks have yielded small amounts of gold, copper, sapphires, rubies, and dark to light green nephrite jade. Also, Precambrian black igneous dikes cut across the granite and schist to form dark stripes on the hills and outcrops. The dikes trend easterly and vary in width from a few feet to 500 feet.

Historically, the Granite Mountains have played an important role in the western migration of the mid-1800s. Westbound settlers had to cross the backbone of the Rocky Mountains. To the south, the high ranges of Colorado were a barrier to the conestoga wagons, and the Montana Rockies were even more formidable. The basins of Wyoming provided a natural avenue across the Rockies, so the Oregon Trail was blazed across the central part of the state, right through the Granite Mountains. Had the Granite Mountains not collapsed around 2 million years ago, there would have been a nearly continuous wall of mountains from the Laramie Range to the Wind Rivers, and the pioneers would have had to find another route leading west!

Wyoming Jade

"Wyoming jade" is not the mineral "jadeite," which is a sodium-rich alumino-silicate mineral found in Burma, Japan, California and Guatemala. Wyoming jade is a variety of the amphibole mineral group called nephrite, a dense, fine-grained variation of actinolite, which contains calcium, magnesium, iron, silica and oxygen. Actinolite is a relatively common mineral in metamorphic rocks, and comes from the non-granitic Precambrian rocks in the area. Iron makes nephrite green and variations in the shade of green depend on the amount of iron in the specimen. Uniformly smooth and apple-green specimens are the most valuable. Veins of nephrite jade occur in the Granite Mountains, but most has been found in conglomerate and boulder deposits of Tertiary age around the flanks of the Granite Mountains. Samples may be purchased at local rock shops.

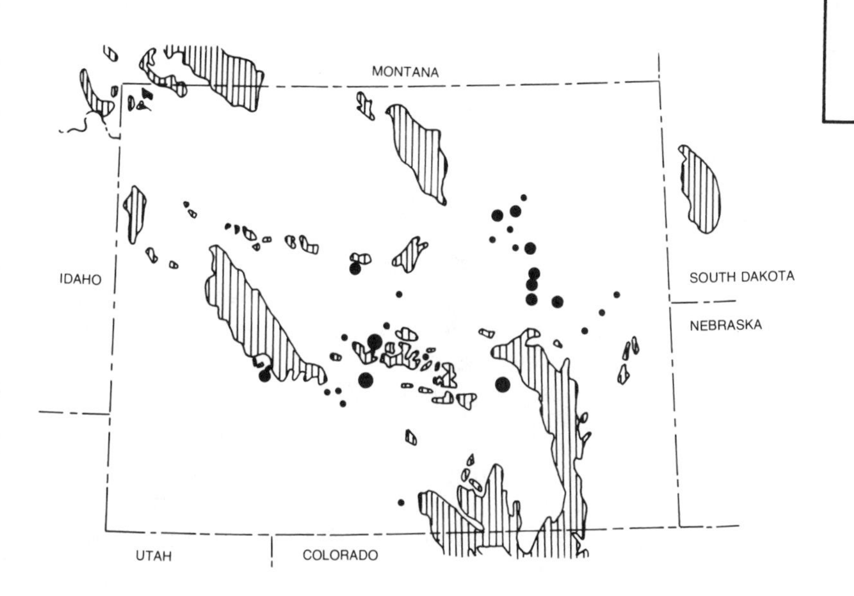

Location of uranium deposits in Tertiary sedimentary rocks of Wyoming. Uplifted areas of Precambrian-age mountain cores are striped pattern. –Adapted from the Geological Survey of Wyoming's 1979 Wyoming Mines and Minerals Map and Public Information Circular #8.

Uranium

The Granite Mountains contain large, economic deposits of uranium. Uranium is an accessory element found in minerals like zircon, tourmaline, mica and feldspar, all of which are common minerals in granitic igneous rocks. It is believed that the source of the uranium could be the granitic bedrocks of the Granite Mountains, or weathered feldspar-rich sediment derived from the granite. Another possible source could be uranium-rich layers of "tuff" that were deposited in lake sediments of Pliocene age, called the Moonstone formation. Tuff is composed of microscopic mineral fragments that were blown from a distant, erupting volcano, then carried by water or wind and deposited as a white, extremely fine-grained sedimentary rock. Tuff contains the same elements as intrusive granitic rocks, including uranium.

Given these possible uranium sources, Mother Nature still had to find a way to concentrate the uranium into bodies of ore. Ground water migrating through fractures and pore spaces in the source rocks picked up uranium and carried it away in

solution. The uranium was later redeposited in porous sandstone layers. For this to occur, the host sandstone had to have sufficient porosity and permeability to carry the uranium-bearing ground water, and the chemical environment of the host rock had to be "reducing," meaning that it lacked free oxygen. Uranium is only deposited in an oxygen-deficient environment. Concentrated uranium minerals were precipitated as microscopic pore-space fillings and as coatings on sand grains in the host sandstone beds.

Two principal uranium mining districts exist in the Granite Mountains — Gas Hills and Crooks Gap. The Gas Hills district, the largest, is on the northern margin of the Granite Mountains, approximately 50 miles southeast of Riverton. A significant percentage of the United State's uranium production has come from this district. The uranium there is in feldspar-rich sandstone beds in the upper part of the Eocene Wind River formation. These sands were deposited as alluvial fans off the rising Granite Mountains during the Laramide orogeny. The uranium was emplaced sometime during Pleistocene time, within the last few million years. The Crooks Gap district is in the southern Granite Mountains, south of Jeffrey City. The Eocene Wasatch and Battle Springs formations are the host strata for uranium mineralization of Crooks Gap. Crooks Mountain, Crooks Gap, and Crook County were all named after General George H. Crook, famous for his Indian fighting exploits.

Wyoming 220
Casper—Rawlins
117 mi./187 km.

The road between Casper and Rawlins crosses several large structures and a wide variety of rocks. At the north end, Wyoming 220 swings around the west end of Casper Mountain, a Precambrian-cored anticline that was shoved north along a thrust fault. The fertile valley at the big bend of the North Platte River is known as "Bessemer Bend." It was the site of the first cabin in Wyoming built by white men, the returning Astorians, in 1812. A town grew up on the

The Hayden Geological Survey party, 1870, west of Casper in the Red Buttes area. This photo is by the famous western photographer, W.H. Jackson, who stands at far right. –Ft. Caspar Museum

site and, in 1889, vied with Casper for the seat of Natrona County. It lost, and has all but disappeared.

Poison Spider Creek flows into the North Platte River from the west at Bessemer Bend. Oil from a seep along the creek was the first sold in Wyoming. In 1851, Jim Bridger, Kit Carson and Cy Iba mixed the oil with flour and sold it to emigrants for axle grease.

At the west end of Casper Mountain, near Goose Egg, bright red sandstones and shales of the Triassic Chugwater and Permian Goose Egg formations crop out. These are ancient shoreline deposits folded into a series of small anticlines. Between here and Alcova Reservoir, the highway follows the North Platte River as it flows through non-resistant Cretaceous black shales. These marine rocks were deposited on the floor of a shallow sea that connected the Gulf of Mexico with the Arctic Ocean about 80 million years ago. Geologists called this ancient sea the "Cretaceous Western Interior Seaway."

Alcova Reservoir was built for flood control and to maintain a regulated water supply for domestic use downstream. Before the construction of Pathfinder and Alcova dams in 1909 and 1938, Casper experienced seasonal flooding along the North Platte floodplains. Alcova Dam and Reservoir are in northeast-dipping Mesozoic and Paleozoic strata and are part of the northeast shoulder of the Granite Mountains uplift. The dam site is in a narrow canyon that the North Platte River cut through the Pennsylvanian Tensleep formation, a sandstone deposited around 300 million years ago. The Tensleep was folded into the Alcova anticline, then offset by the reverse fault on its southwest flank. The main body of the reservoir lies over an adjacent broad syncline. Similarly, the dam site at Pathfinder Reservoir is in a

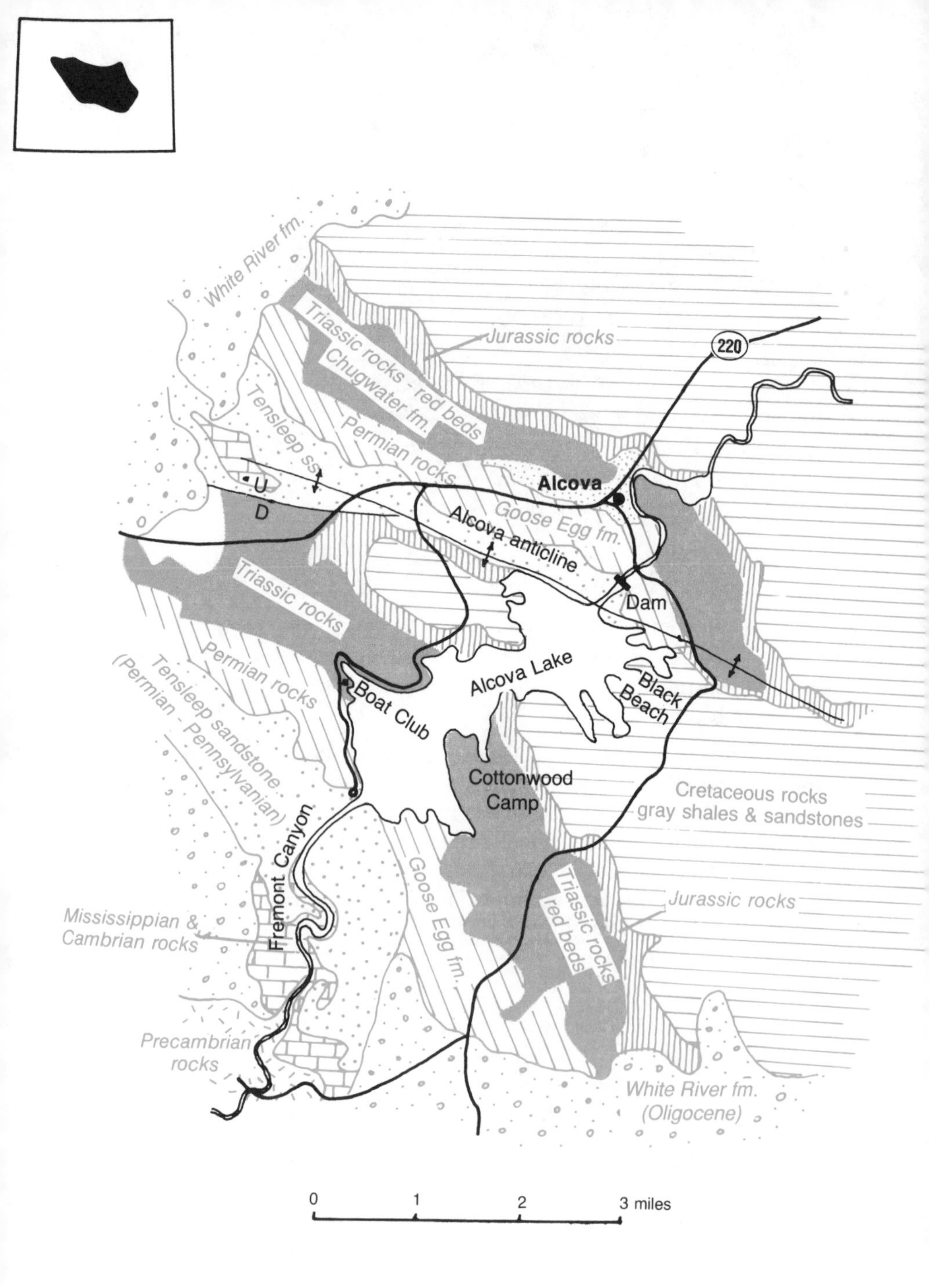

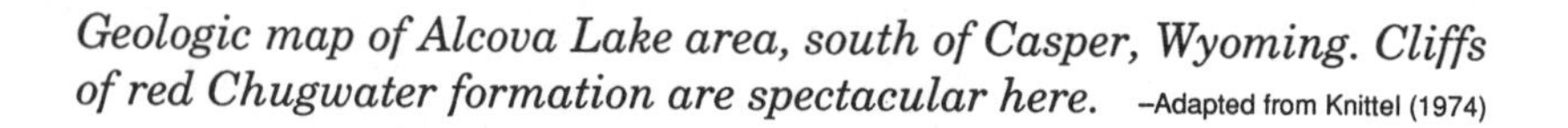

Geologic map of Alcova Lake area, south of Casper, Wyoming. Cliffs of red Chugwater formation are spectacular here. –Adapted from Knittel (1974)

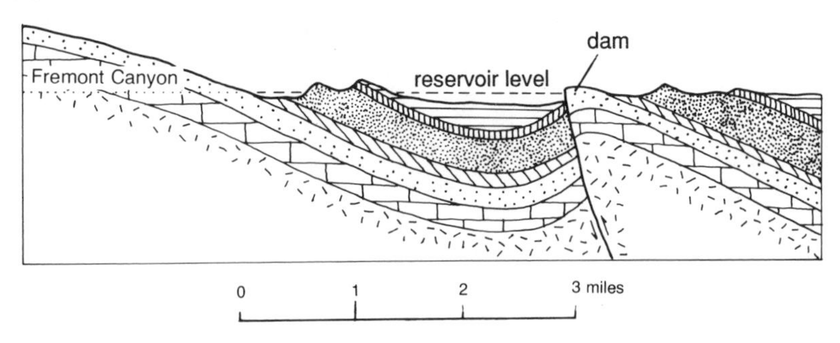

Cross section Alcova Reservoir. For key to rock symbols, refer to geologic map on facing page. –Adapted from Knittel (1974)

narrow canyon cut by the North Platte River through Precambrian basement rocks.

Between Alcova and Muddy Gap, Wyoming 220 crosses the central Granite Mountains graben, a trough dropped along faults at the north and south flanks of the Granite Mountains over the last two million years. Precambrian granitic outcrops, like Independence Rock, rise above a thin veneer of Miocene sedimentary rocks, mostly white, soft, tuffaceous sandstones of the Splitrock formation.

Independence Rock

South of Alcova, Wyoming 220 passes Independence Rock, an important historical site. It was a landmark for explorers, fur trappers, emigrants, Forty-Niners, cavalry, stage coaches, and Pony Express riders. It has been called the "Great Register of the Desert" because of

Independence Rock, photographed by W.H. Jackson for the Hayden Geological Survey in 1870. The view today is essentially the same. –Ft. Caspar Museum

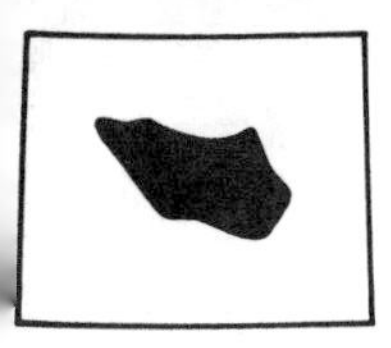

Summit of Independence Rock along Highway 220 between Muddy Gap and Alcova Reservoir. The rounded surface is due to exfoliation of granite.

the many names carved in its granite surface. The following quote is from a passerby, William Marshall Anderson, in 1834:

> We have breakfasted this morning at the base of Rock Independence. There are few places better known or more interesting to the mountaineer than this huge boulder . . . On the side of the rock names, dates and messages, written in buffalo grease and powder, are read and re-read with as much eagerness as if they were letters . . . from long absent friends . . . being a place of advertisement, or a kind of trapper's post office . . . It is a large, egg-shaped mass of granite, entirely separate and apart from all . . . ranges of hills. One mile in circumference, about six or seven hundred feet high, without a particle of vegetation . . .

Various stories surround the origin of the name Independence Rock. The mountain man "Broken-Hand" Fitzpatrick may have named it on July 4th, 1824, when his buffalo hide "bull-boat" capsized near the rock. General Ashley camped here on July 4th, 1825, and may have named it. Others say that Captain Bonneville named it in

Etched name at Independence Rock, "The Register of the Desert" on Highway 220, 53 miles southwest of Casper

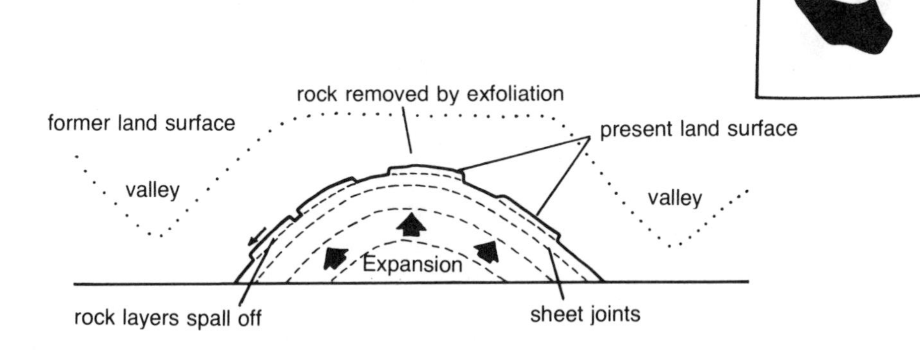

Independence Rock shows how exfoliation, caused by pressure release and rock expansion, can form rounded domes.

1832 because it stands alone and "independent" on the prairie. Pioneers on the Oregon Trail needed to make it to the rock by July 4th to complete their journey west before winter, and therefore called it Independence Rock.

Wagon ruts of the Oregon Trail are still visible southwest of the rock. The present solitude and quiet of Independence Rock belies its history. During the heyday of the Oregon Trail in the 1840s to 1850s, thousands of covered wagons were scattered over the surrounding prairie each night during mid-summer. A sizeable town eventually rose by the rock, but the wind, prairie grass and rattlesnakes have recaptured everything. Today, Independence Rock has returned to the solitude it enjoyed before 1830. Please recognize the historical value of this unique landmark and do not mark its surface!

Independence Rock is the product of a surficial geological process called exfoliation. When rocks are buried underground, they are under confining pressure from all sides, including the weight of overlying rock. When rocks are exposed to the surface by erosion, they are no longer under confining pressure from above, and therefore expand outwardly along curved fractures that spall off the surface like the layers of an onion. Layers of granite have broken off, one after the other, to form the rounded mass of rock shaped like a turtle shell. Similar round monoliths of granite throughout the central Granite Mountains formed in the same way. Wind-blown sand and silt have grooved the rock and polished it to a high gloss in places, a process called wind faceting. Independence Rock is most certainly not the result of glacial erosion and polishing, as is commonly purported.

Wyoming 220 parallels the course of the Sweetwater River south of Independence Rock. To the west another geological landmark, Devil's

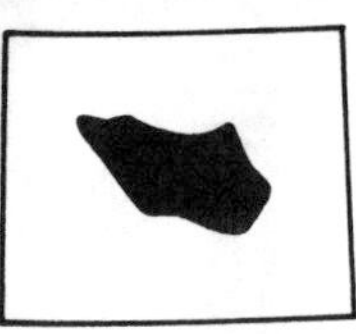

Gate, is visible. At this point, the Sweetwater River makes an abrupt turn and flows through a granite ridge, cutting a chasm 330 feet deep. It is 400 feet wide at the top, but only 30 feet wide at the bottom.

Muddy Gap is a narrow canyon through the Ferris Mountains at the south flank of the Granite Mountains. The Ferris Mountains are composed of up-turned Precambrian, Paleozoic and Mesozoic rocks that were pushed south along the south flank of the Granite Mountains during the Laramide orogeny. Very young normal faults on the north side of the Ferris Mountains dropped the central part of the Granite Mountains, preserving the unique, sediment-covered landscape.

Lamont and Bairoil are old Wyoming oil towns. Several oil fields exist in this area, but the Lost Soldier field west of Bairoil is king. It was discovered in 1916 and has since produced more than 200,000,000 barrels of oil and 76,000,000 cubic feet of gas from many different horizons, principally the Pennsylvanian Tensleep sandstone, Mississippian Madison limestone, and Cambrian Flathead sandstone.

The long stretch of flat highway between Lamont and Rawlins crosses Quaternary sand and silt deposits that cover the Cretaceous Cody black shale. The structure and stratigraphy of the Rawlins uplift are discussed in the Interstate 80 section from Laramie to Rawlins.

US 287
Wyoming 789
Muddy Gap—Lander
82 mi./131 km.

This route crosses two distinct sections divided by the Beaver Rim. On the east, the road traverses the west end of the Granite Mountains and follows the course of the Sweetwater River. West of Beaver Rim, the road cuts through Mesozoic strata folded into a series of anticlines on the northeastern flank of the Wind River Mountains.

Another famous Oregon Trail landmark is Split Rock, 8 miles west of Muddy Gap, to the north. Its summit elevation is 7,305 feet. The notch resembles a gun sight when viewed from either the east or west and is visible for more than 50 miles. Rapid erosion along a joint or

Split rock between Muddy Gap and Jeffrey City along Highway 287.

fracture system that trends east-west opened the notch. Water slowly worked its way into the fractures and weathered the granite. Water carried away the debris, forming the deep cleft in the top of the mountain.

Along the eastern section of Wyoming 287 through the Granite Mountains, you see outcrops of pink granitic basement, like Split Rock, north of the highway. Crooks and Green mountains lie to the south. Green Mountain is composed of resistant boulder conglomerates belonging to the Eocene Crooks Gap formation, which was shed off the south flank of the Granite Mountains while the Rocky Mountains formed. Crooks Mountain is composed of up-turned and resistant Paleozoic and Mesozoic rocks, locally overlain by Eocene conglomerate.

Jeffrey City is in the middle of the Crooks Gap uranium mining district. Wyoming's first uranium mill was built here in 1957. Before that time, Jeffrey City was only a post office known as "Home on the Range." This is also the center of the Wyoming jade area. A single chunk of jade weighing 3,366 pounds was supposedly found near here in 1943!

The Ice Slough, a natural phenomenon with historical interest, is mid-way between Jeffrey City and Sweetwater Station. Ice was uniquely preserved here during the summer months beneath a thick, insulating cover of grasses and swampy soil. Mountain men and emigrants used the ice to cool their meat, and to enjoy an occasional whiskey on the rocks.

The Beaver Rim is a spectacular escarpment that overlooks the Wind River Basin to the west and north. It is composed of a nearly complete section of Tertiary sedimentary formations, a classic area

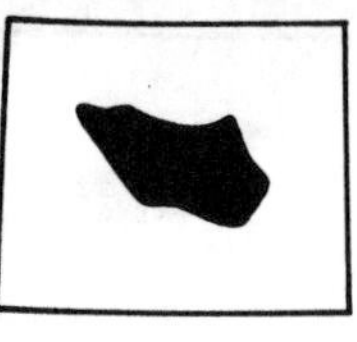

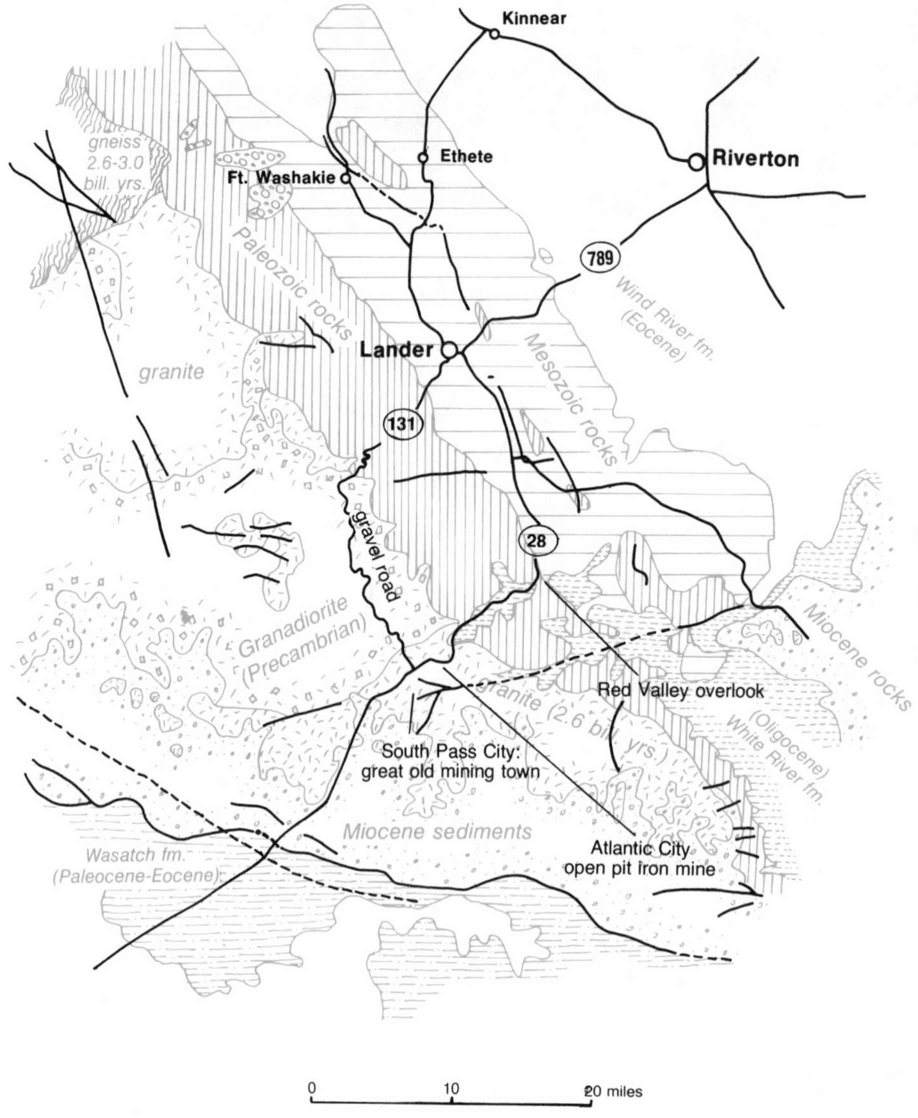

Geologic map of South Pass-Lander area, southern Wind River Mountains.

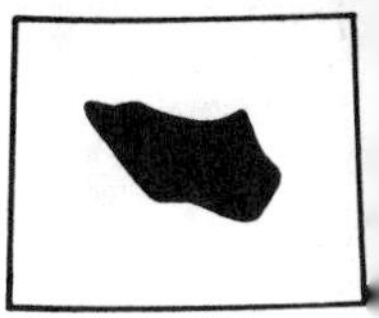

White River formation unconformable over Cretaceous shales 15 miles southeast of Lander on US 287.

for studying the last 65 million years of Wyoming history. In descending order, the Tertiary section is composed of:

1) the Miocene Splitrock formation (soft tuffaceous sandstone);

2) the Oligocene White River formation (brilliantly white tuffaceous claystone and sandstone);

3) the Eocene Wagon Bed formation (green and gray tuffaceous claystone, sandstone, and conglomerate);

4) and the Eocene Wind River formation (variegated claystone and sandstone).

All of these Tertiary units unconformably overlie northeast-dipping, gray to black Cretaceous shales and sandstone.

US 287 cuts through the Derby Dome oil field about 5 miles east of the intersection with Wyoming 28. Derby Dome is a small anticline with oil production from the Permian Phosphoria and Pennsylvanian Tensleep formations. As of 1985, a total of 1,365,947 barrels of oil had been produced. The red shales and sandstones in the middle of the fold belong to the Triassic Chugwater formation. North of the intersection of Wyoming 28 and US 287 the road crosses the black Cretaceous Cody shale into Lander, passing by Dallas Dome oil field, where Wyoming's first commercial oil well was drilled in 1884.

Wyoming 28
Lander—Farson: South Pass
77 mi./123 km.

Wyoming 28 traverses the southern end of the Wind River Mountains through South Pass. The Continental Divide trends southeast off the crest of the Wind River Range to South Pass, then splits to

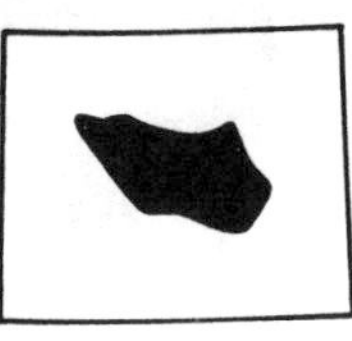

Cross-bedded, wind-blown sands of the Jurassic Nugget sandstone on the "Red Grade" along Highway 28 south of Lander.

encircle the Great Divide Basin. South Pass does not resemble what one would imagine as a typical point along the Continental Divide, and that is why it is so special. It is a gentle rise reaching only 7,526 feet. South Pass was the "hump" of the Oregon Trail, the open door to the west—here the Continental Divide is easily crossed.

Robert Stuart's Astorians may have come through South Pass on their return from the Northwest in October, 1812, but the effective discovery of the pass was made in March of 1824 by Jedediah Smith, famed explorer and member of the fur trapping partnership Smith-Jackson-Sublette. This was a critical moment in American history—knowledge of this pass opened up the west and the flood of emigration began.

Geologically, the road between Lander and Farson may be divided into three segments: 1) the northeast-dipping "backside" of the Wind River Range from Lander to Atlantic City; 2) the crest of the southern Wind River Range between Atlantic City and the Continental Divide; and 3) the Green River Basin. We discuss each segment below and, depending on which way you are traveling, you can reverse the order of the paragraphs.

Wyoming 28 passes the Dallas Dome oil field about 6 miles southeast of Lander. A "great tar spring" was first reported in this area by Captain Bonneville in 1827, an early hint of the vast petroleum potential. Dallas Dome was the site of the first oil commercial well in

Panoramic view of the "Red Grade" from Red Canyon Overlook on Highway 28 south of Lander. The valley has been cut into the Triassic Chugwater formation, while the more resistant Jurassic Nugget sandstone forms the cliff on the mountains.

Wyoming, drilled in 1884 by Mike Murphy. This was the discovery well for the Dallas Dome oil field, which has since produced over 10,000,000 barrels of oil from the Chugwater, Phosphoria, and Tensleep formations over its 100 year history. The field is still pumping oil and produced 116,460 barrels of oil from 59 wells in 1985. Anticlines such as Dallas Dome are fairly common on the flanks of Wyoming's mountains, and were major targets for early oilwell drillers. Today, the "easy oil" has been found and geologists must use sophisticated seismic techniques to explore the deeper parts of basins.

South of the intersection, Wyoming 28 follows soft, black, marine shales and sandstones of Cretaceous age, then turns southwest and begins a gradual climb over the southern Wind River Mountains. Some of the best outcrops of Paleozoic and Mesozoic strata anywhere in the state are along this beautiful section. In descending stratigraphic order to the southwest, you will see:

1) the pink to orange, cross-bedded, Jurassic Nugget sandstone and bright red Triassic Chugwater formation, forming spectacular ridges along the "Red Grade;"

2) the long, grass-covered slopes of the Permian Phosphoria formation;

3) the gray Mississippian Madison limestone;

4) the white dolomite cliffs of the Ordovician Bighorn formation;

5) and the Cambrian Gallatin limestone, Gros Ventre shale, and the Flathead sandstone.

The scenic "Red Grade" was a wagon road from Lander to Atlantic City and was so-named because of the long climb from the valley to uplands along Red Canyon Creek. The Red Grade is an excellent example of differential erosion of sedimentary layers: the harder, more resistant sandstone layers, like the Nugget formation, form cliffs and benches, whereas the soft, easily eroded shales and siltstones of the lower Chugwater form valleys and gullies.

The crest of the southern Wind River Range is crossed between Atlantic City and the Continental Divide. Miocene rocks, about 27 million years old, locally overlie Precambrian basement rocks. The Miocene rocks are siliceous and arkosic sandstones, claystones and conglomerate, whereas the underlying Precambrian basement is composed of metamorphosed sedimentary rocks that are over three billion years old. They include metamorphosed sandstone and con-

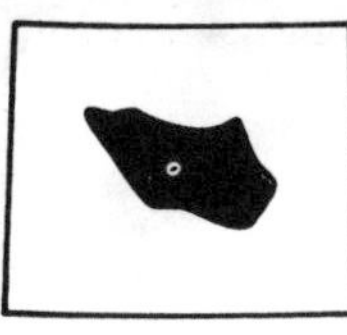

South Pass City Mine south end of Wind River Mountains on Highway 28.

glomerates, mica schists, graphite schists, metamorphosed iron formation, and black amphibolite gneisses. These Precambrian rocks started as muddy sediments and basalt deposited under water along the margins of an ancient North American continent, and then were subjected to very high temperatures and pressures during a mountain building event around 3 billion years ago. The rocks were deeply eroded until about 570 million years ago when they were buried under the first Paleozoic marine sediments spread across Wyoming.

Atlantic City and South Pass City

Gold nuggets were discovered in the clear streams near South Pass in 1842. Although prospectors worked the area for many years, the gold rush came in 1867 with rich discoveries at several locations. Miners flocked to the area by the hundreds and three gold camps were established — South Pass City, Atlantic City, and Miner's Delight. By 1871, over 2,000 people lived in the camps, and twelve gold-processing stamp mills were in operation. But as quickly as it came, the boom died, apparently because of mine flooding and labor disputes. Many of the miners went south to work in coal mines along the Union Pacific Railroad. South Pass City eventually became an Overland Stage and Pony Express station. Old mines, equipment, and these colorful towns are only a few miles south of Wyoming 28 on an excellent gravel road. The side trip is worth it!

In 1868, Esther Hobart Morris made history by hosting a tea party in South Pass City for the Republican and Democratic candidates for the Territorial Legislature. She made them pledge that if elected, they would support and work for women's suffrage. Bright, the Republican, won the election and introduced a bill adopted by the first legislature of the Territory of Wyoming giving women full rights to vote and hold office, making Wyoming the first governmental unit in the world to do so! When Wyoming achieved statehood in 1890, it earned the nickname "Equality State," by including women's suffrage in its constitution. Esther Morris later served as Justice of the

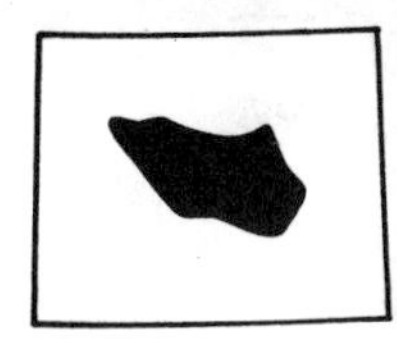

Banded iron formation at Atlantic City Mine on Highway 28 southwest of Lander.

Peace of South Pass City, the first female judge in the world. It is ironic that women's suffrage was championed in the "Wild West" where "men were men," but it reflects the independent attitude and open-minded quality of western people, as well as the high value of women in frontier communities.

In recent years, iron ore has replaced gold as the primary resource in the South Pass area. Deposits of "banded iron formation" in ancient Precambrian rocks were strip-mined by U.S. Steel Corporation from 1962 to 1983 at their Atlantic City mine. The open pit is easily seen from the highway. Semi-refined pellets of concentrated iron ore called "taconite" were produced at South Pass and shipped to Utah for further processing. The layered iron formation was probably deposited as a chemical precipitate with other sediments on a seafloor more than 3 billion years ago, then was thickened by complex folding during an ancient, Precambrian orogeny.

The road crosses the northeast corner of the Green River Basin between the Continental Divide and Farson, a distance of about 35 miles. The road crosses South Pass then follows Pacific Creek, which drains from Pacific Springs on the west side of the Continental Divide. Discovered by mountain man Sublette in the 1820s, the spring offered Oregon Trail pioneers their first taste of water flowing west to the Pacific Ocean.

Miocene sandstones and claystones overlie Precambrian metamorphic rocks from the Continental Divide southwest for about 4 miles. The road crosses the Wind River thrust fault about 5 miles southwest of the Continental Divide, but it is not visible because it is buried under Eocene strata of the Wasatch formation. For the next 13 miles, the road crosses the main body of the Wasatch formation, drab sandstone, variegated claystone and siltstone, along with locally derived gravel near the mountains. The sediments were shed off the

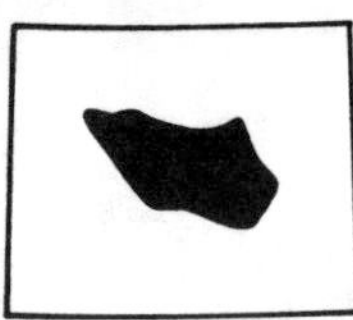

rising Wind River Range as the Rocky Mountains formed. The road crosses the Laney member of the Eocene Green River formation for the last 17 miles into Farson. It is about 45 million years old and consists of brown to tan oil shale and soft limestones. Fossil fish abound.

EASTERN WIND RIVER BASIN

US 20/26
Casper—Shoshoni—Riverton
98 mi./157 km. 22 mi./35 km.

Casper to Hell's Half Acre

Many Wyomingites regard the drive between Casper and Shoshoni as the "longest drive in the state." Low, rolling, sagebrush-covered hills, not much to see on the horizon, and the sheer distance from anywhere to anywhere probably contributes to this sentiment. But the geology, albeit somewhat subtle, is indeed interesting. Reading the landscape dispels the normal boredom of the drive.

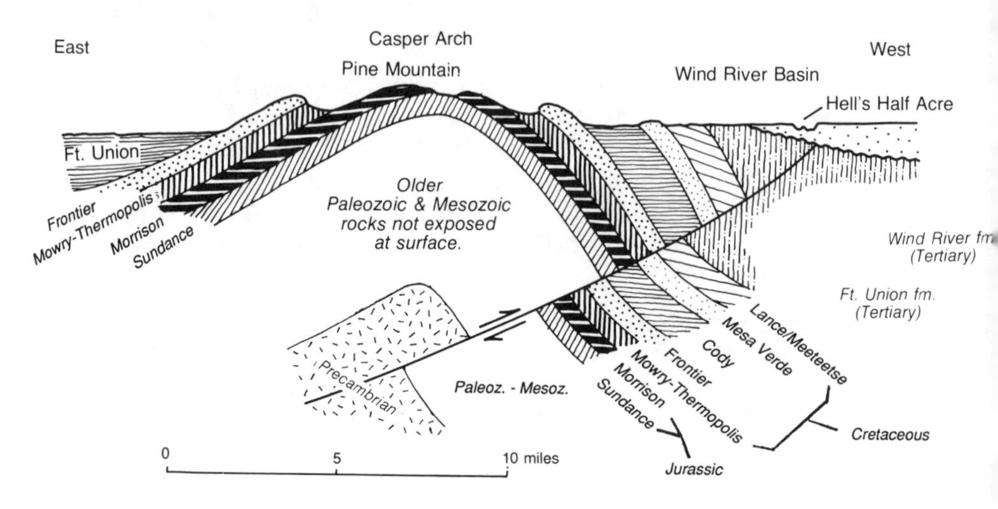

Cross section of Pine Mountains and Hell's Half Acre, looking south from US Highway 20-26, near the town of Powder River.

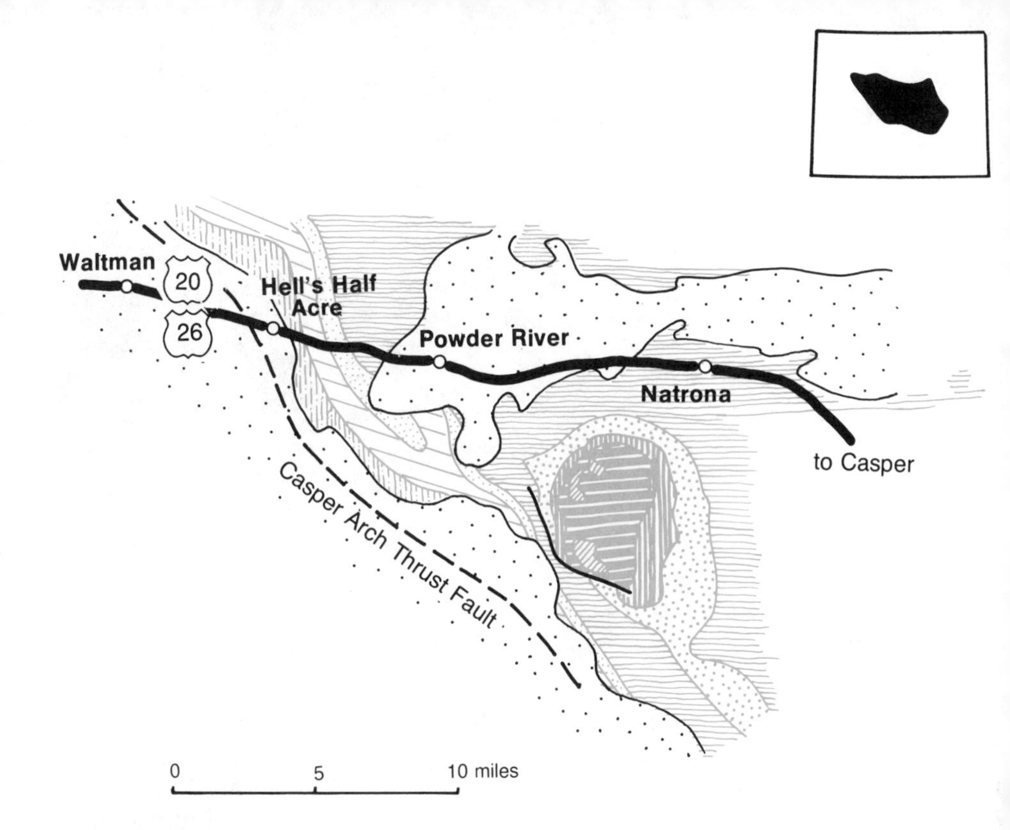

Geologic map of Pine Mountain and Hell's Half Acre. Symbols are same as cross section.

At the east end of this highway segment, the Emigrant Gap anticline appears to the south where steeply tilted slabs of Frontier sandstone outline the fold. Between Casper and Hell's Half Acre, the road crosses Cretaceous strata on the Casper arch, a broad, thrust faulted, anticlinal uplift between the Wind River and Powder River basins. It is also a sag or saddle between the Laramie Range to the southeast and the Bighorn Mountains to the northeast, a section that was not uplifted as much. The arch is highly asymmetrical, steep and thrust faulted on its western flank but gently east-dipping on its eastern flank. The Cretaceous Cody shale crops out over much of the arch, but it is easily eroded and does not form good outcrops. The Cody shale consists mostly of dull-gray shale, with some gray siltstone and fine-grained gray sandstone. It weathers into surfaces with a distinctive "popcorn" texture because it contains a clay mineral called montmorillonite or bentonite, which alternately swells and contracts with wetting and drying. Occasional outcrops of Cody shale are seen between Casper and Powder River. Younger Tertiary rocks that once covered the Casper arch have been eroded to expose the Cody shale.

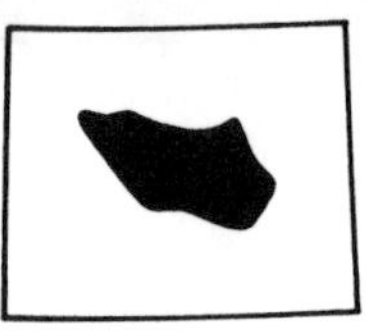

Pine Mountain Dome structure, from the air along US 20, US 26, 25 miles west of Casper.

White caliche soils are seen in low areas where water stands in winter and evaporates in summer, leaving a residue of salts and calcium carbonate.

Although not topographically high or rugged like most uplifts, the Casper arch has been spectacularly thrust faulted to the west, forming what petroleum geologists call a large "mountain-front overthrust." The eastern Wind River Basin extends several miles eastward beneath the Casper arch, forming a "subthrust block." These subthrust blocks are very important to petroleum geologists because they contain sedimentary rocks that may generate and trap oil or natural gas. One of the first major subthrust discoveries in the Rockies was made on the Casper arch in 1980, when an oil company drilled a well through 8,835 feet of Precambrian granite in the hanging wall of the Casper arch, and then drilled to almost 20,000 feet in subthrust Mesozoic strata. The well was completed for gas in the Frontier formation. This discovery was important because it proved that thick stratigraphic sections may extend under the edges of mountain ranges, and may contain oil and gas.

The Casper arch is a funnel or corridor for the prevailing westerly winds, similar to the Leucite Hills in the Green River Basin. These strong winds formed the sand dunes that lie along a west-to-east trend from the Wind River Basin across the central Casper arch. Many of these dunes are now dormant and covered with grasses, but others still move east. The dune field probably formed between 12,000 and 19,000 years ago during or immediately following the retreat of Pinedale glaciers of about 70,000 to 15,000 years ago. The road cuts through many dune deposits on the west side of the arch near Powder River. Look for the loose sand.

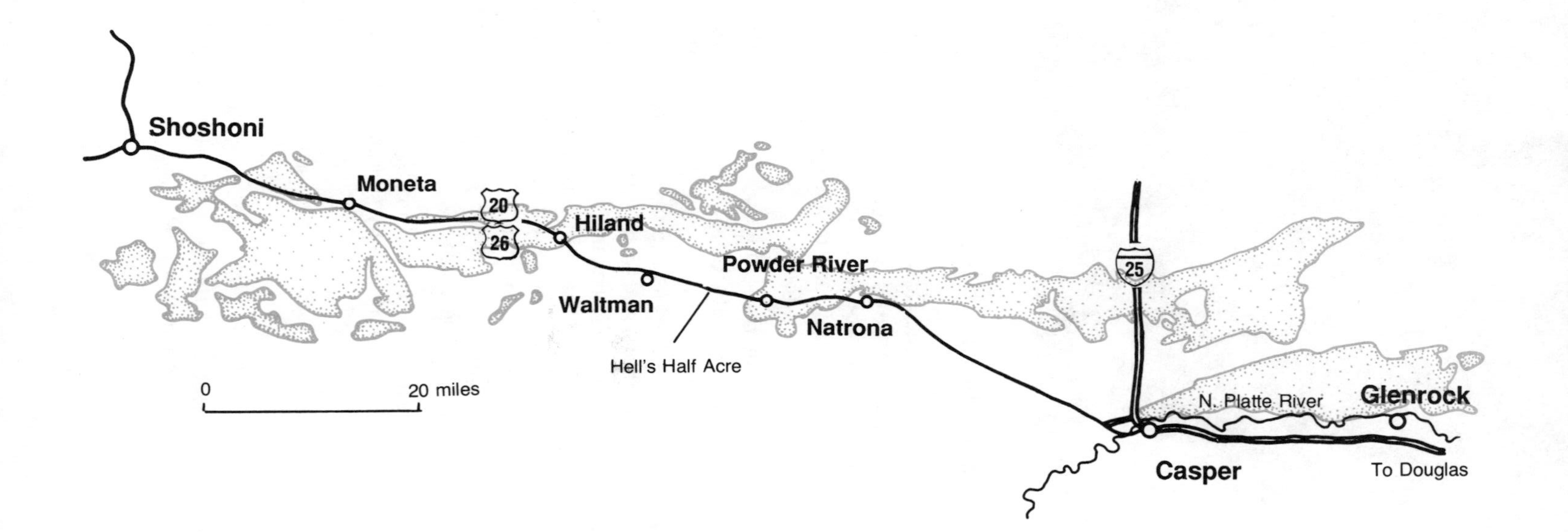

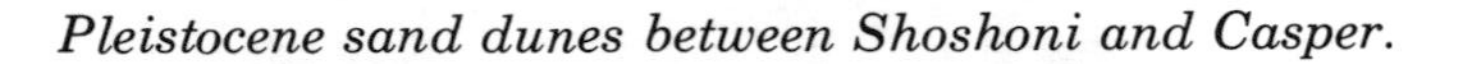

Pleistocene sand dunes between Shoshoni and Casper.

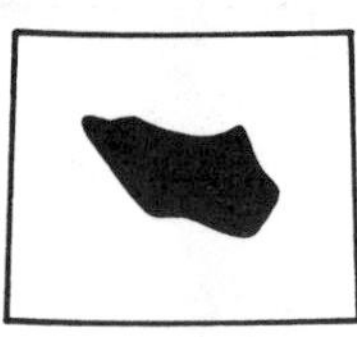

Hell's Half Acre between Casper and Shoshoni on Highway 26-20. Hell's Half Acre is a badlands carved from upper Cretaceous, Paleocene and Eocene sands, shales, and coals.

Hell's Half Acre—Shoshoni—Riverton

Hell's Half Acre marks the western side of the Casper arch. This is a spectacular example of wind and water acting on soft bedrock to produce badlands topography. The site was once known as Devil's Kitchen, but was changed to Hell's Half Acre by a promoter who used that name on postcards.

Hell's Half Acre is carved out of three formations. In the deep, eastern parts of the chasm the upper Cretaceous Lance and Paleocene Fort Union formations dip very steeply to the west. The Lance is buff sandstones and greenish shales, whereas the Fort Union contains brown and gray sandstone, gray shale, and thin coal beds. These units are overlain on the western side of the chasm by the Eocene Wind River formation, which dips gently to the west into the Wind River Basin. It consists of red, pink and yellow claystone and sandstone. The contact between the Wind River and underlying units is an important key to the geologic history of the area. The following events produced the landscape you see today:

1) The upper Cretaceous Lance and Paleocene Fort Union formations were deposited around 65 million years ago by streams or in flood-plains and swamps.

2) The Casper arch was raised during the Laramide orogeny, about 55 to 60 million years ago. It was thrust to the west and the Lance and Fort Union were folded to near-vertical beds.

3) Erosion bevelled the Casper arch, truncating the Lance and Fort Union formations to produce an erosion surface.

4) The Wind River formation was deposited by streams around 50 million years ago. These strata were flat at the time of deposition.

5) Continued uplift of the Casper arch, or subsidence of the Wind River Basin, tilted the Wind River formation westward; the underlying strata also rotated more steeply to the west.

6) The four formations have been re-exposed by erosion over the last few million years. Water from sporadic rain showers and snowmelt slowly carved into the soft sandstones and claystones, producing the badlands topography of the chasm. Wind carried some of the finer, eroded silt and sand away to the east, but the small amount of yearly precipitation and runoff has been responsible for most of the erosion.

Between Hell's Half Acre and Shoshoni, the road crosses the basin proper and Tertiary strata in the footwall of the Casper arch thrust. It is a featureless landscape on the Eocene Wind River formation, locally overlain by recent sand dune deposits. Stream deposits of the Wind River formation consist of red, pink, and yellow claystone and sandstone. The relatively flat ground surface of the basin has remained essentially unchanged for many thousands of years. The Owl Creek Mountains frame the horizon to the north. Several tributaries of the Wind River drain this part of the basin, including Badwater Creek, Alkali Creek and Poison Creek. Obviously, the surface water is a bit "hard," to say the least!

"Hoodoos," strongly eroded outcrops that take on human-like forms, appear a few miles east of Shoshoni. Nearby, where the usually dry channel of Poison Creek is near the highway, look for eroded sand dunes that formed during the ice age from river-bottom sand that was blown to the east.

US 26 follows the Wind River between Shoshoni and Riverton, as it flows across the basin to Boysen Reservoir. Buttes near Shoshoni are capped by orange colored sandstones, underlain by variegated claystones of the Eocene Wind River formation.

WESTERN WIND RIVER BASIN

The western and northern part of the Wind River Basin is framed by the northeastern flank of the Wind River Mountains and the Owl Creek Mountains, respectively. This is a beautiful and variable part of the state—the alkali and sage flats of the basin floor contrast

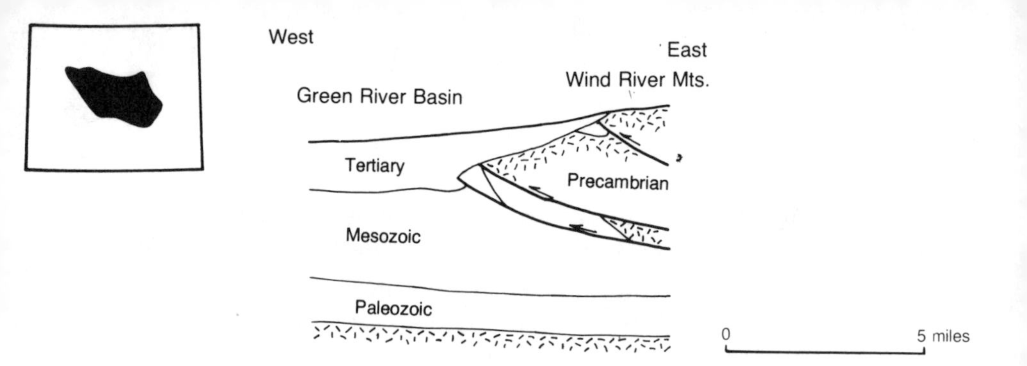

Cross section of Wind River Mountain front. –Adapted from Gries (1983)

sharply with the alpine meadows and snow-capped peaks near Togwotee Pass.

Wind River Mountains

The Wind River Mountains are one of the most awesome collection of peaks anywhere in the Rocky Mountains. The high, glacier carved peaks rival any mountains in Colorado or Montana for their ruggedness, isolation, and spectacular beauty! A snowflake falling on the 13,000 foot crest of the "Windies" can melt down the southwest flank into the Pacific Ocean via the Green and Colorado rivers, or down the northeast flank into the Gulf of Mexico via the Wind-Bighorn River, Yellowstone River, and Missouri-Mississippi rivers. This is literally the backbone of the continent! The Indians called it "the top of the world."

The Wind River Range contains fifteen peaks over 12,000 feet, culminating in Wyoming's highest point, Gannett Peak at 13,785 feet. This glacier-covered peak was named for Henry Gannett, a geographer and member of the Hayden Geological Survey in the 1870s. The Dinwoody Glaciers, the largest ice field in the U.S. outside of Alaska, rests at the high northern end of the range. The Bridger Wilderness area encompasses most of the range.

The Windies are more than 100 miles long by about 40 miles wide. The range is a coherent block of Precambrian crust bound on its southwest flank by one of the largest thrust faults in the Rockies. During the Laramide orogeny, 60 to 55 million years ago, the range was raised and thrust over 13 miles to the southwest over the thick sedimentary deposits of the Green River Basin. The Wind River thrust fault dips northeast at an angle of about 30 degrees beneath the range and has been traced on reflection seismic records to a depth of at least 15 miles into the crust! Following the Laramide orogeny, through much of Tertiary time, the Wind River Mountains were

eroded and partially covered with sedimentary deposits. The nearly flat erosional surface is well exposed on the southwest flank of the range north of Pinedale. During the last two million years, the buried mountains and basins have been exhumed by stream erosion, and the mountains were glaciated at least three times to produce the present landscape.

Wind River Indian Reservation

The Wind River Indian Reservation occupies a large part of the western Wind River Basin. This, the only Indian reservation in Wyoming, is headquartered at Fort Washakie. The reservation is populated by two great Indian tribes, the Shoshone and Arapahoe.

The Shoshonean territory initially extended from the Cascades to the Rockies, south almost to Mexico. Later, because the Shoshone were not warlike, they were forced to live in the Rockies for protection. The Sheep Eaters of Yellowstone Park, a band of the Northern Shoshone, were driven into the remote mountains. The Shoshone acquired horses by 1740, probably by trading with the Commanches to the south. During the mid-1800s, the Northern Shoshone tribe, under the leadership of the great Chief Washakie, made their home in the Bridger Valley of southwestern Wyoming.

The Arapahoe tribe migrated onto the Great Plains from the upper midwest, perhaps from as far as eastern Canada. The name Arapahoe was corrupted by Lewis and Clark from the Pawnee named "Laripihu," meaning "he who buys or sells;" the word "Arapahoe" does not exist in the Arapahoe language! The Arapahoe were at the geographic center of a large Plains Indian confederacy in the 18th and 19th centuries that included the Gros Ventre, Blackfeet, Sioux, Kiowa, Taos Pueblo, and Jicarilla Apache tribes. In the great peace treaty signed at Fort Laramie in 1851, the Arapahoe and Cheyenne tribes were allocated a corridor of land southward from the North Platte to the Arkansas River, and east from the Rockies to Kansas, "for as long as the grass shall grow." With the discovery of gold at Pikes Peak in Colorado in 1858, this treaty was quickly overlooked by whites.

The Wind River Indian Reservation was allocated to the Shoshones by treaty with Chief Washakie on July 3, 1868, at Fort Bridger. The government wanted the Shoshone Indians out of their native Bridger

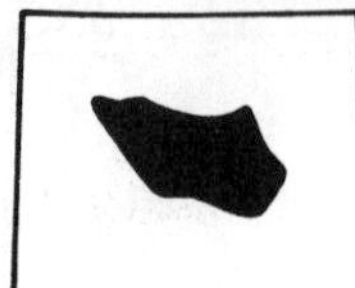

Valley in southwestern Wyoming to make room for the Union Pacific Railroad. Then, on March 18, 1878, the U.S. Army escorted 938 starving Arapahoe Indians onto the Shoshone Reservation, indifferent to the traditional enmity between the Shoshone and Arapahoe. Chief Washakie repeatedly asked the government to move the Arapahoe off their deeded land, but nothing was done. The Shoshone tribe sued the government for giving part of their land to the Arapahoe and eventually won $4,408,444 in 1938.

Although the size of the reservation has dwindled considerably, it now contains 2 million acres and is home to around 4000 people of the Arapahoe and Shoshone tribes. Visitors wishing to hike, hunt, or fish on the reservation must obtain permission and special permits.

US 287
Riverton/Lander—Dubois —Moran Junction
55 mi./88 km. 75 mi./120 km.

Riverton is an agricultural, oil and mineral center for the Wind River Basin, named for the confluence of the Wind, Little Wind and Popo Agie rivers. The road between Riverton and Moran Junction follows the northeast side of the Wind River Mountains, crossing over the north end of the range at Togwotee Pass. US 287 passes through the heart of the Wind River Indian Reservation. Geographically and geologically, this route may be subdivided into three sections: 1) Riverton or Lander to the intersection of US 287 and /US 26; 2) the US 287/US 26 intersection to Dubois; and 3) Dubois over Togwotee Pass to Moran Junction.

US 26
Riverton—US 287/26 Intersection

US 26 crosses a plateau on the Eocene Wind River formation between Riverton and the intersection, overlooking the broad valley of the Wind River and the Wind River Mountains to the southwest.

Pilot Butte, a knob of rock north of the highway near Pilot Butte Reservoir, is composed of resistant sandstones of the Eocene Wind River formation. US 26 crosses the Pilot Butte oil field, located on an

anticline five miles east of the intersection; this field has produced over 14,000,000 barrels of oil since its discovery in 1916. The broad arching or "roll-over" of this anticline is visible south of the highway.

Lander—US 287/26 Intersection

Lander is located by the Popo Agie River (pronounced pō-pō'-zha, which is Crow for "head water"), along a well traveled Indian and game trail. The "Sinks" of the Popo Agie River, an interesting natural site, are 10 miles south of Lander, on Wyoming 131 at Sinks Canyon State Park. Here, the river sinks into a cavern in the Madison limestone and flows underground for about 600 feet to the "rise," where it reappears in a crystal-clear, trout-filled pool.

Lander was originally known as "Push Root" because spring winds seemed to push the new plants up by their roots. The local Indians called the area the "valley of warm winds," although many other places in Wyoming sport much stronger winds! The town was renamed in 1869 for General F.W. Lander, who was sent west in 1857 to inspect the Oregon Trail; he established the "Lander Cut-off" of the Oregon Trail in the Green River Basin.

The Lander oil field is immediately northeast of town on Wyoming 789 to Riverton. A well exposed anticline, the Lander field was discovered in 1909, and has produced over 11,000,000 barrels of oil from the Permian Phosphoria and Pennsylvanian Tensleep formation. This is one of several anticlines along the gentle, back side of the Wind River Mountains.

Between Lander and Fort Washakie, the road follows a broad valley cut into soft, east-dipping, black shales of the Cretaceous Cody formation. These are marine shales deposited at the bottom of a shallow seaway about 80 million years ago.

Between Fort Washakie and the intersection of US 287 and US 26, the road straddles the Cretaceous Frontier formation, gray sandstone and sandy shale, on the east side of the road; and the Eocene Wind River formation, variegated claystone and sandstone with lenses of conglomerate, on the west side of the road. The Frontier formation was steeply folded on the west limb of the Sage Creek anticline, which has produced oil from the Phosphoria and Tensleep formations. The Winkleman Dome oil field just north of Sage Creek, about 9 miles from Fort Washakie, has produced about 83,000,000 barrels of oil

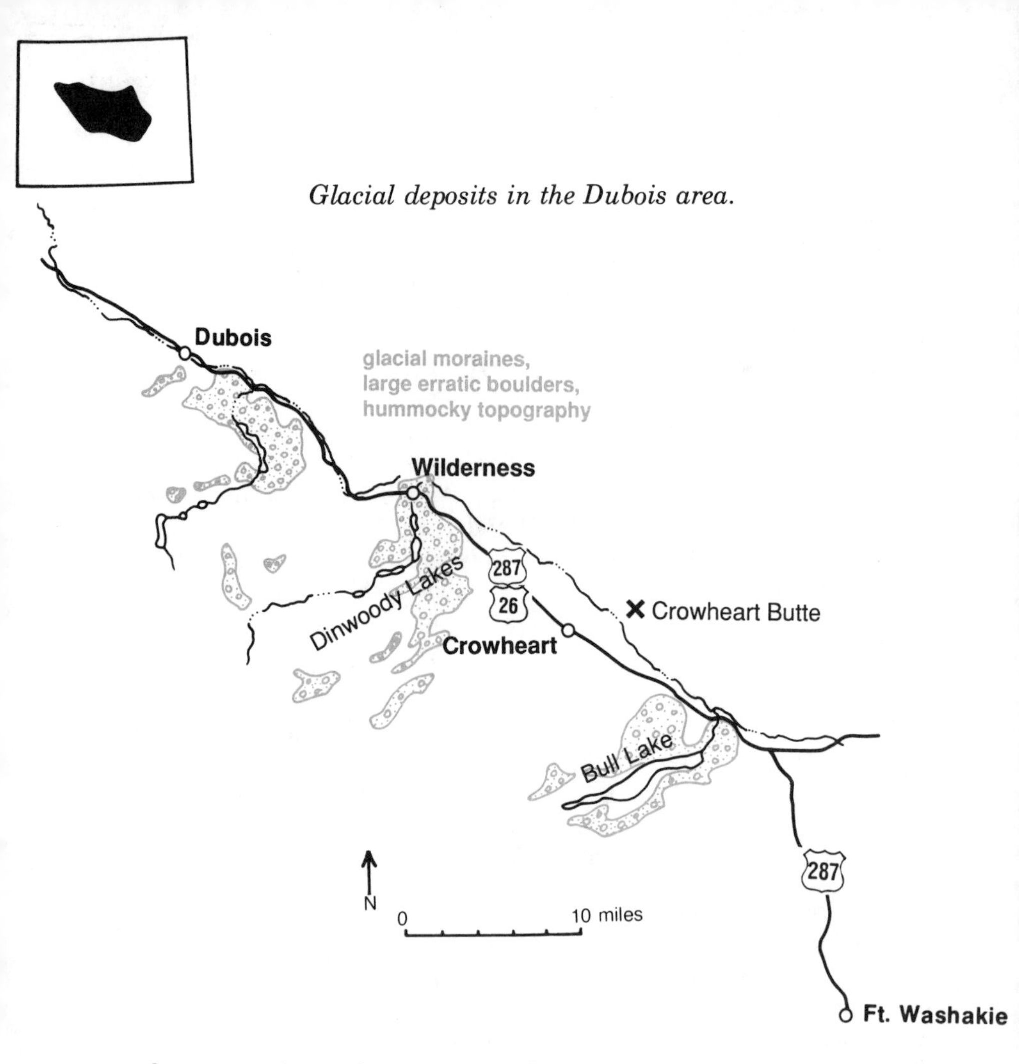

Glacial deposits in the Dubois area.

from a variety of Mesozoic and upper Paleozoic horizons; it was discovered in 1917. The large surface anticlines in the Wind River and Bighorn basins are everywhere marked by bright red shales of the Triassic Chugwater formation.

US 287/26 Intersection to Dubois

The road follows the main channel of the Wind River from the intersection of US 287 and US 26 to Dubois. This narrow arm of the Wind River Basin is framed to the west by the slopes of the Wind River Mountains, and to the north by folded and faulted Paleozoic and Mesozoic strata overlain by Eocene volcanic rocks of the Owl Creek and Absaroka ranges. The basin floor is composed of the Eocene Wind River formation, stream-deposited, claystone in various colors and

rusty-brown feldspar-rich sandstone.

For a short side-trip, take the gravel road to Bull Lake, a classic site for studying the recent glacial history of Wyoming. Downstream from the dam, along Bull Lake Creek, are several recessional moraines left as glacial ice melted back into the high mountains. These younger moraines were deposited by the "Pinedale" glaciation between 70,000 and 15,000 years ago, and display well-developed "knob-and-basin" topography. The boulders at the surface are almost unweathered. Scattered deposits from the older "Bull Lake" glaciation, 200,000 to 130,000 years ago, are on the flats above the valley. These moraines are more subdued and the boulders more weathered. This area is the type locality of the Rocky Mountain Bull Lake glacial event. Fremont Lake, near Pinedale on the west side of the Wind River Mountains, is the namesake for the Pinedale glacial deposits.

Crowheart Butte, the isolated butte north of the highway, is composed of claystone and sandstone of the Eocene Wind River formation. Behind its name is a colorful story. In 1866, Indian tribes in the area fought a battle to establish supremacy over hunting grounds in the basin. Chief Washakie led the Shoshone and Bannock tribes against the Crow Indians, led by Chief Big Robber. In an effort to save lives, Chief Washakie suggested that he and Chief Big Robber fight alone at the top of this butte—the winner would eat the other's heart! Washakie won and the butte was named "Crow Heart." In his old age, Washakie was asked if he actually ate Big Robber's heart; he replied "youth does foolish things."

The big bend in the Wind River, 15 miles northwest of Crowheart Butte, offers a spectacular view of almost barren Jurassic and Cretaceous strata, unconformably overlain by the Eocene Wind River formation. The section dips northeast from the blood-red sands and shales of the Triassic Chugwater formation, through the Jurassic Sundance formation, to purple and black shales of Cretaceous age. The overlying, horizontal Wind River formation forms spectacular badlands that display bright red and white strata between here and Dubois.

Dubois nestles in a valley that the Wind River cut into soft Triassic, Jurassic, and Cretaceous strata, overlain by the Eocene Indian Meadows formation. Red and orange dip slopes of the Triassic Chugwater formation are exposed on the southwest side of the Wind River. The hand-dug cave on the west end of town is in the Jurassic Nugget

sandstone; it was used in the past as a bar and jail.

Dubois was originally called "Never Sweat," a name apparently derived from early, lazy residents of the area. In 1886, a new post office was built, but the Postal Service refused to accept the name Never Sweat. Instead, they selected the name Dubois for Senator Dubois of Idaho, who was on the Senate postal committee at the time.

Between Dubois and Togwotee Pass the road follows colorful outcrops of the Eocene Wind River formation, which has weathered into a badlands topography. Outcrops of Mississippian, Pennsylvanian, and Permian age strata slope off the Wind River Range on the southwest side of the highway. Towering peaks of the Absaroka volcanics rise on the skyline to the west.

The old Union Pass road intersects US 287/26 about nine miles north of Dubois. Union Pass is 9,210 feet high and was named because it unites the Atlantic and Pacific watersheds. This point on the Continental Divide marks a division in the waters of three of the continent's great rivers, the Missouri/Mississippi, Columbia, and Colorado. Indians traveled through the pass to reach hunting grounds in Jackson Hole long before white men entered the west. These trails were used by John Colter in 1807, by fur trappers, prospectors, hunters, and Hayden's geological survey party in the 1870s.

Across from Union Pass Road, the broad Du Noir River Valley joins the Wind River Valley. The Du Noir River cut its gorge in soft, colorful strata of the Eocene Wind River formation. Then a Pleistocene glacier gouged the gorge out into a wider valley. Ramshorn Peak on the north skyline is composed of horizontal layers of volcanic conglomerates and ash of the late Eocene Wiggins formation.

East of Togwotee Pass, Brooks Lake Creek cuts a beautiful little gorge into soft strata of the Eocene Aycross formation, which overlies the Wind River formation. It is a lower unit of the Absaroka volcanic pile. Aycross strata are brightly colored claystones and tuffaceous sandstones that were deposited on the margins of the Absaroka Mountains volcanic field to the north, around 49 million years ago. Patched and diverted road sections show that they are very subject to slumps and landslides.

The Absaroka volcanic field extends north from Togwotee Pass into the Gallatin Range of Montana. It was formed by repeated eruptions

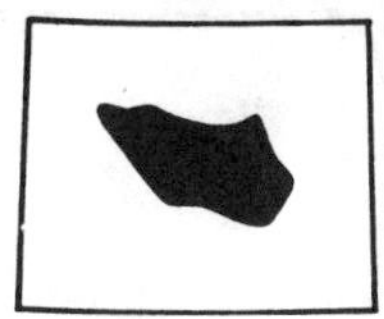

of andesite and dacite in mid-Eocene time. The rock types are like those of the Cascade volcanoes in Washington and Oregon, so this area may have looked similar to the Cascades when the volcanoes were active. Be sure to stop at Falls Campground and take a short hike along the rim of the falls; note that the waterfall at the head of the gorge is held up by the resistant ledge of rock.

Pinnacle Buttes form high, rugged peaks on the north side of the highway, east of Brooks Lake. These peaks are carved from the Eocene Wiggins formation, which is about 46 million years old and contains conglomerate composed of pebbles of volcanic rocks, along with interbedded layers of white volcanic ash. Flat-topped Lava Mountain, on the south side of the highway, is composed of nearly 1,000 feet of horizontal lava flows of black basalt that are estimated to be about 500,000 years old.

Togwotee Pass, on the Continental Divide, is at 9,544 feet. Togwotee was a feared medicine man of the Sheep Eater tribe of the Northern Shoshone, and was a sub-chief under Chief Washakie. The Sheep Eaters lived in the high country in and around Yellowstone Park. The following quote describes the life style of the tribe:

> They stayed up there in the mountains. They did not go among the Plains Indian buffalo eaters. They used dogs for packing and watching their pack horses. They used snow shoes and could run and jump between cliffs with these. It was a hard life in the mountains. In the fall they would come down to the foot of the mountains. They did not like to dance or anything like that, they just looked for their food. They were clean people.
>
> from *The Wind River Reservation Yesterday and Today*

At Togwotee Pass, the Wiggins formation makes up the horizontally layered gray volcanic conglomerates and white tuffs on Sublette Peak to the northeast and Two Ocean Mountain to the southwest. The Eocene Aycross formation forms the greenish, treeless badlands of claystone southwest of Togwotee Pass.

On the high, west side of Togwotee Pass, above Togwotee Mountain Lodge, the road cuts through Mesozoic and upper Paleozoic strata on the south flank of the Washakie Range. The core of the Washakie Range contains basement rock that was shoved west on the Buffalo Fork thrust fault, then covered beneath thousands of feet of volcanic debris from the Eocene Absaroka volcanic field. The Mississippian

Madison limestone, through the Cretaceous Frontier formation, form steeply dipping slopes on the north side of the highway. Togwotee Pass descends to the west into Jackson Hole. About nine miles west of the summit, the Togwotee scenic overlook offers a spectacular panoramic view of the Tetons and Jackson Hole.

The lower part of the west side of Togwotee Pass, below Togwotee Mountain Lodge to Moran, cuts through the upper Cretaceous Harebell formation, about 5,000 feet of gold-bearing quartzite conglomerates, olive-drab sandstone, and green claystone. This formation slumps readily, so road damage is common.

The source of the quartzite conglomerates in the Harebell formation has been an interesting dispute among geologists. Some geologists have suggested that they were eroded from a raised block of basement rock northwest of the Teton Range that was later buried by volcanic rocks. Others believe these quartzite gravels were carried into Wyoming from Idaho on thrust faults. The final verdict is still not in.

The floodplain of the Buffalo Fork River forms the valley floor east of Moran Junction. Hills north and south of the valley contain more Cretaceous sandstones, shales, and conglomerates.

Road guides north and south of Moran Junction are found in the Grand Teton chapter.

NORTH WIND RIVER BASIN

The northern Wind River Basin lies between the Owl Creek Mountains and the Bridger Range. The Wind River Basin is deepest along its northern margin adjacent to the Owl Creek Mountains; in some places the basin holds as much as 30,000 feet of sedimentary strata from the ground surface to the top of Precambrian basement rock below. Several major oil and gas fields produce from these rocks along the northern margin of the basin. The Madden field near Lysite produced 327,000 barrels of oil and 220 billion cubic feet of gas from the Fort Union and Mesaverde formations between 1969 and 1985.

Owl Creek—Bridger Mountains

The Owl Creek—Bridger mountains are one of three, major east-

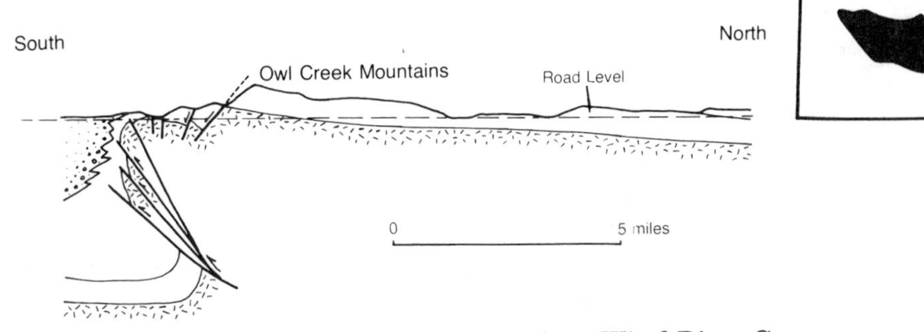

Cross section of Owl Creek Mountains along Wind River Canyon.
–Adapted from Wise (1963)

west trending ranges in the central Rocky Mountains. The other two are the Uinta Mountains in Utah and the Granite Mountains in central Wyoming. All the other ranges trend north or northwest, suggesting they experienced east-west compression and shortening during the Laramide orogeny. The east trend of the Uinta Range is explained by ancient, east-striking faults that were easily reactivated to raise the range. The Granite and Owl Creek mountains are bounded by thrust faults, along their south margins, that shoved these ranges southward over the adjacent Green River and Wind River basins, respectively. However, the west ends of these thrust faults curve to the north-northwest parallel to the grain of other Laramide faults and uplifts, so they may still be explained by east-west compression of the crust during the Laramide.

The Owl Creek Mountains are a highly asymmetrical anticline, meaning that strata on opposite sides of the range are bulged up at different angles. Paleozoic and Mesozoic strata on the south flank dip very steeply southward, whereas strata on the north flank dip gently northward at about ten degrees. The core of the range is ancient Precambrian metamorphic and igneous rock, probably around three billion years old.

The Owl Creek Mountains are structurally more complex than just a simple asymmetric uplift that was thrust to the south. They merge on the east into the northern Casper arch by the way of a large thrust

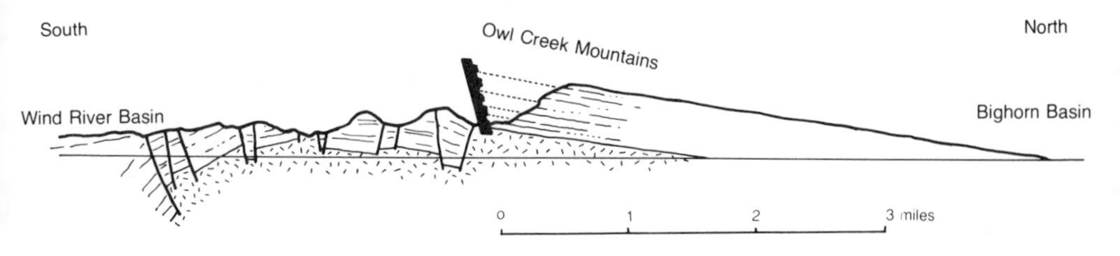

Cross section of Wind River Canyon. –Adapted from Wise (1963)

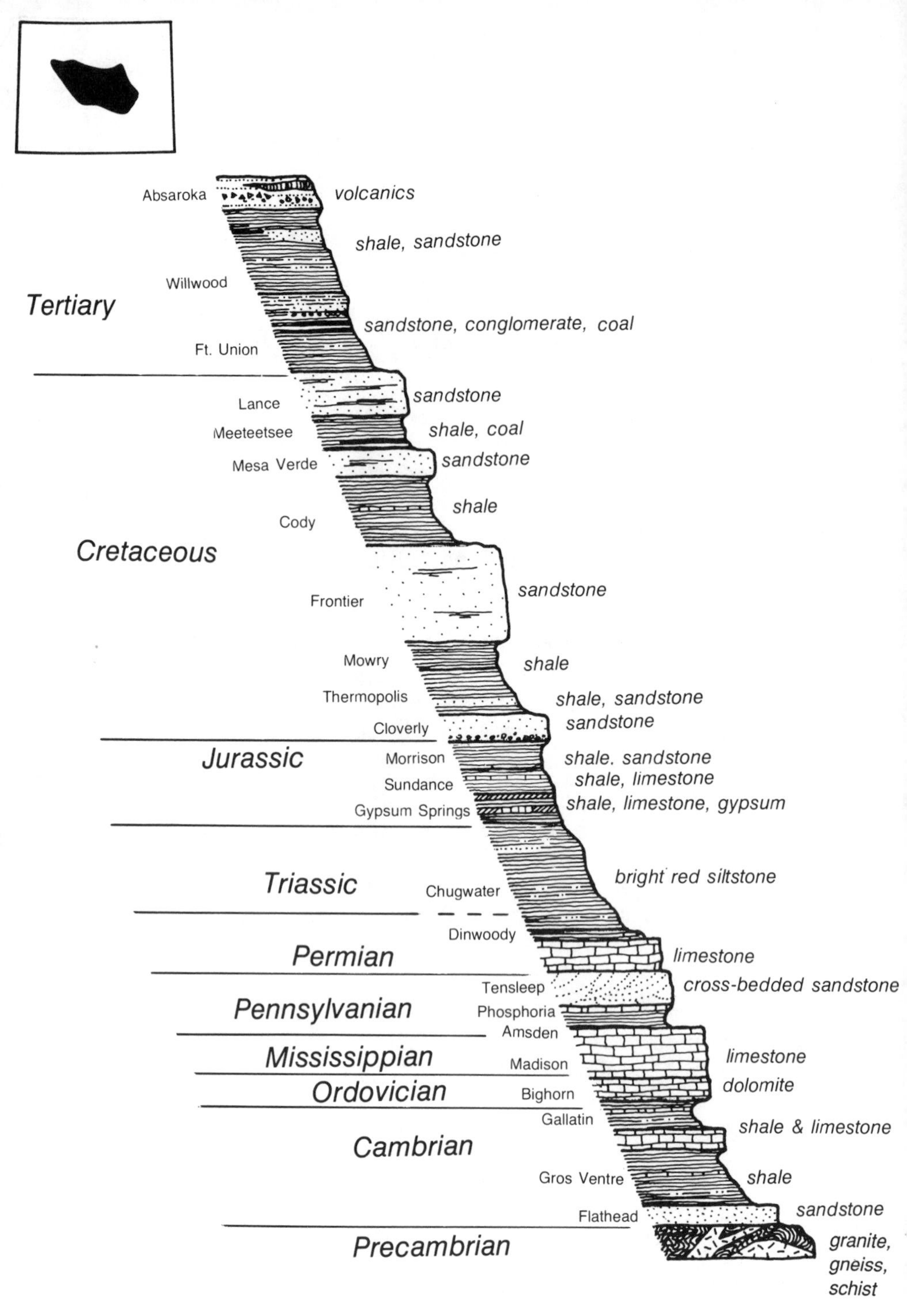

Stratigraphic column in Wind River Canyon.

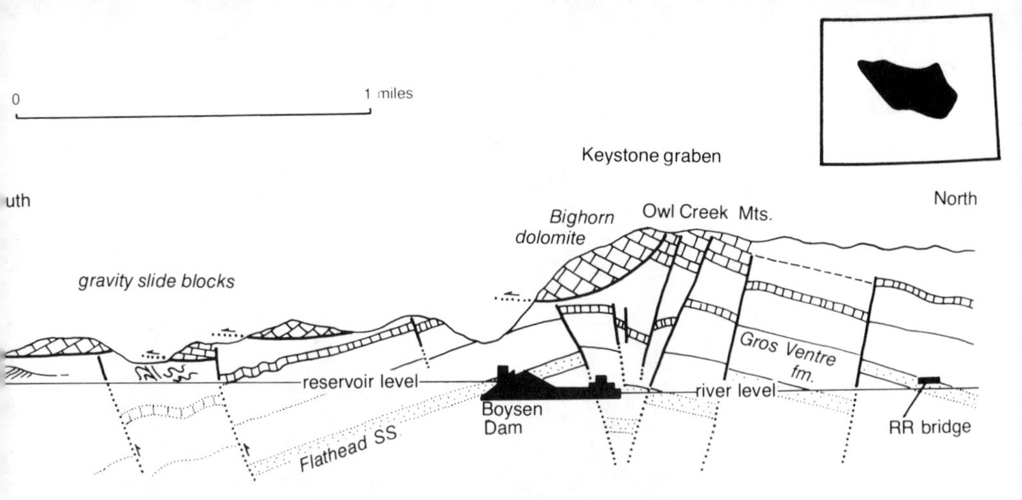

Cross section of Keystone Graben and Gravity Slides at Boysen Dam in Wind River Canyon. –Adapted from Wise (1963)

fault that extends along the west side of the arch. This thrust is discussed in the section from Casper to Shoshoni. In contrast, the Owl Creek Mountains west of Wind River Canyon are a series of faulted blocks of Precambrian basement and overlying starta that trend northwest and form a staggered series of anticlines and synclines. In addition, the south flank of the range has been broken up by a series of high-angle normal faults that define a central "keystone graben." Furthermore, blocks of Paleozoic limestone and dolomite detached and slid downhill to the south over other sedimentary strata. The net result is a mountain range that stands out only because of its east-west trend, but also because of the complicated faulting along its south margin.

US 20, Wyoming 789
Shoshoni—Thermopolis:
Wind River Canyon
32 mi./51 km.

The Wind River Canyon offers one of the best exposed and complete sections of sedimentary rocks in central Wyoming. The canyon has been cut by the Wind River through the backbone of the Owl Creek Mountains. As the Wind River emerges from the north end of the canyon, it becomes the Bighorn River and flows north through the Bighorn Basin and the northern Bighorn Mountains, eventually to join the Yellowstone River in Montana.

Normal fault that offsets Cambrian strata at south end of Wind River Canyon along Highway 20.

Why did the Wind River cut this canyon? Why didn't the river continue to flow east to join the North Platte River? This is another example of an illogical river in the Rocky Mountains—it appears to have followed the most difficult course possible. The answer lies in a process called "superposition," which also occurred on the Hoback, Snake, Green, and North Platte rivers. After the Laramide orogeny, the Rocky Mountain uplifts were eroded for millions of years during the middle and late Tertiary. Because of the dry climate of the time, and the lack of through-flowing streams, much of the eroded sediment was deposited in adjacent Laramide basins, like the Wind River and Bighorn basins. By late Tertiary time, the combination of erosion in the mountains and deposition in the basins had produced a nearly flat topography over Wyoming. Streams established their courses at this time, totally indifferent to the fact that they were flowing over buried uplifts of Precambrian rock and strata. In the last 2 million years, the climate has become wetter and the Rocky Mountains were regionally uplifted; this caused the streams to downcut their channels. If a section of channel was over a buried uplift, like the Owl Creek Range, the river cut into the underlying rocks and eventually formed a canyon. In this manner, the Laramide uplifts and basins of

Boysen normal fault at south end of Wind River Canyon along Highway 20. Dark Precambrian metamorphic and igneous rocks on the right have been up-thrown relative to Paleozoic strata to the left.

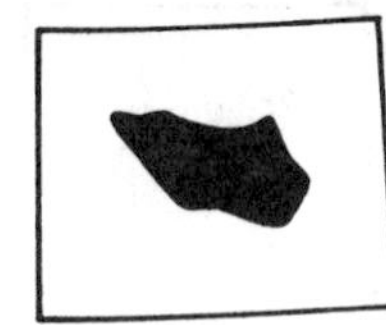

Coarse-grained, light-colored igneous "pegmatite" veins of quartz and feldspar in Precambrian rocks at south end of Wind River Canyon along Highway 20.

the Rocky Mountains have been "exhumed," or dug out by the major river systems and their tributaries. They have cut deep canyons in the process. Therefore, the seemingly illogical courses of many Rocky Mountain rivers are due to the fact that their courses were determined before the present landscape of ranges and basins existed.

Between Shoshoni and the south canyon entrance, US 20 follows the east side of Boysen Reservoir over the Eocene Wind River formation. This unit is composed of variously colored claystone and sandstone with some lenses of conglomerate all deposited by ancient streams flowing from the Owl Creek Mountains. About 1 to 2 miles south of Boysen Dam, is a region of complex faulting formed by "roll-over" or arching of the south flank of the Owl Creek Mountains. The dam is near the top of this arch, as indicated by the change from south-dipping strata south of the dam to north-dipping strata north of the dam.

The canyon narrows considerably as you cross the Boysen normal fault and pass into uplifted, dark, Precambrian metamorphic rocks. Three tunnels are cut through these Precambrian rocks. Road signs

Wind River Canyon

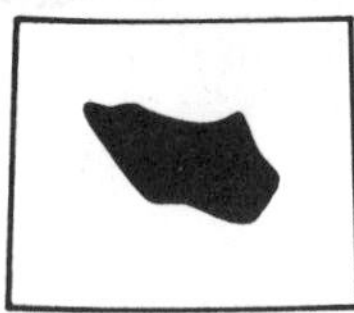

Cross-bedded wind-blown Tensleep sandstone of Pennsylvanian age at north end of Wind River Canyon along Highway 20.

for a geological tour of the canyon start at a pull-off area just north of the tunnels. Spectacular pink and white "pegmatite veins" of quartz and feldspar cut the darkest metamorphic rock. The light-colored, well-bedded Cambrian Flathead sandstone rests directly on the Precambrian rock—an age difference of about 2 billion years! The Flathead sandstone was the first of many marine sedimentary layers to be deposited over the Precambrian "basement" during the Paleozoic era.

North of the tunnels, the Paleozoic section dips uniformly northward at about ten degrees, forming the spectacular cliffs of the Wind River Canyon. The cliff-forming formations are resistant layers of limestone, dolomite and sandstone, and are identified by signs adjacent to the highway. Soft shales and siltstones form the gentle slopes between the cliffs. Proceeding northbound from the tunnels, the major cliff-formers are the Precambrian metamorphic rocks and the overlying Flathead sandstone, the Ordovician Bighorn dolomite, the Mississippian Madison limestone, and the Pennsylvanian Tensleep sandstone. All are nicely labeled. It is not hard to see that the river cut this canyon one block at a time! Large boulders scattered on the lower slopes and in the river channel were dislodged from the overly-

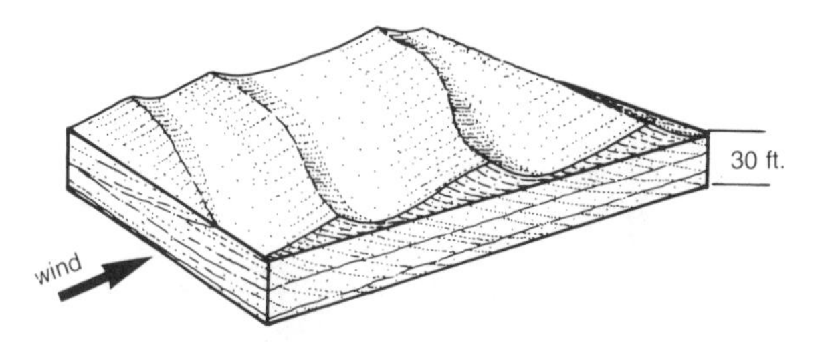

Large scale cross-bedding produced by blowing sand dunes as seen in Tensleep formation at north end of Wind River Canyon.

Hoodoos in Eocene Wind River formation just east of Shoshoni, Highway 20-26.

ing cliffs by the slow processes of weathering and gravity. Vertical fractures aided weathering by providing avenues for water, which chemically and mechanically breaks the rock apart.

The north end of the canyon is marked by blood-red sandstones and shales of the Triassic Chugwater formation. The red color is caused by iron-oxide minerals that glue the microscopic sand grains together.

The town of Thermopolis and its famous hot springs are north of Wind River Canyon. Roadlogs from Thermopolis north are in the Bighorn Basin chapter.

V
Northern Wyoming—The Bighorn Basin

INTRODUCTION

The Bighorn Basin is a large, oval depression in the north-central part of Wyoming. It covers around 10,000 square miles, 120 miles long, by about 60 miles wide. It is bounded on the west by the Beartooth, Absaroka, and Washakie mountains, to the south by the Owl Creek and Bridger mountains, and to the east by the Bighorn and Pryor mountains. The basin is topographically open at its north end, where it merges into the Crazy Mountain Basin of south-central Montana.

The Bighorn Basin offers the special contrast of landscapes and environments that typifies Wyoming. The center of the basin is low, flat, and dry, but the surrounding mountains contain thick forests and rise over 13,000 feet to perpetually snow-capped summits. The four major rivers that drain these mountains and flow across the basin, the Bighorn, Shoshoni, Greybull, and Clarks Fork, eventually empty into the Yellowstone River in Montana. The Bighorn River cuts a deep, narrow canyon across the northern Bighorn Mountains.

The Crow Indians claimed the basin as their hunting grounds. Immense herds of buffalo were native to the basin, as well as deer, elk and antelope. The name Bighorn came from the abundant mountain sheep that inhabited the rocky ledges above the Bighorn River in the early days. In 1807, John Colter, a member of the Lewis and Clark Expedition, was the first white man known to have passed through the area. The basin served as a hub for fur trappers throughout the early

1800s. It was surrounded by rich beaver streams on every side, and provided the first segment of the route to St. Louis—the Bighorn River. Cattlemen did not move into the basin until the late 1870s. The basin had good grazing and was well-watered, but the rugged encircling mountains and great distance from the railroad kept this part of the state unsettled.

Geologically, the Bighorn Basin is often used as a model for interpreting other parts of the Rocky Mountains. It contains well-exposed anticlines and other features produced by the Laramide orogeny, and is a natural "classroom in the field" for geologists and laymen alike.

The Bighorn Basin may be subdivided into three broad zones, ranging from the surrounding mountains to the basin center. The first is the rim of high mountains that surrounds the basin. These mountains were thrust over the basin during the Laramide orogeny, 60 to 55 million years ago. On the west, a major thrust fault along the eastern margin of the Beartooth Mountains and possibly a "phantom" thrust fault south of the Beartooths, uplifted Precambrian basement and overlying strata that project beneath the thick pile of Eocene volcanic rocks of the Absaroka Mountains. On the east, the Bighorn Mountains were thrust westward over the basin margin at the north and south ends of the range.

The second zone is the "shoulder," a platform of anticlines and synclines that encircles the basin. The shoulder is between 5 and 10 miles wide and forms a bench between the main mountain masses and deep trough of the basin. As the mountains rose and the basin subsided during the Laramide orogeny, vertical displacement created a zone of faults that produced the shoulder. It contains many anticlines, some of which trapped oil. The anticlines are typically asymmetric, with their steep limbs facing the mountains; invariably, they are bound by reverse or thrust faults at depth that have uplifted the Precambrian basement, folding the overlying strata in the process. These anticlines are usually identified by the bright red Triassic Chugwater formation, which is exposed in the core or along the flanks of the folds. A classic example is Sheep Mountain anticline just north of Greybull.

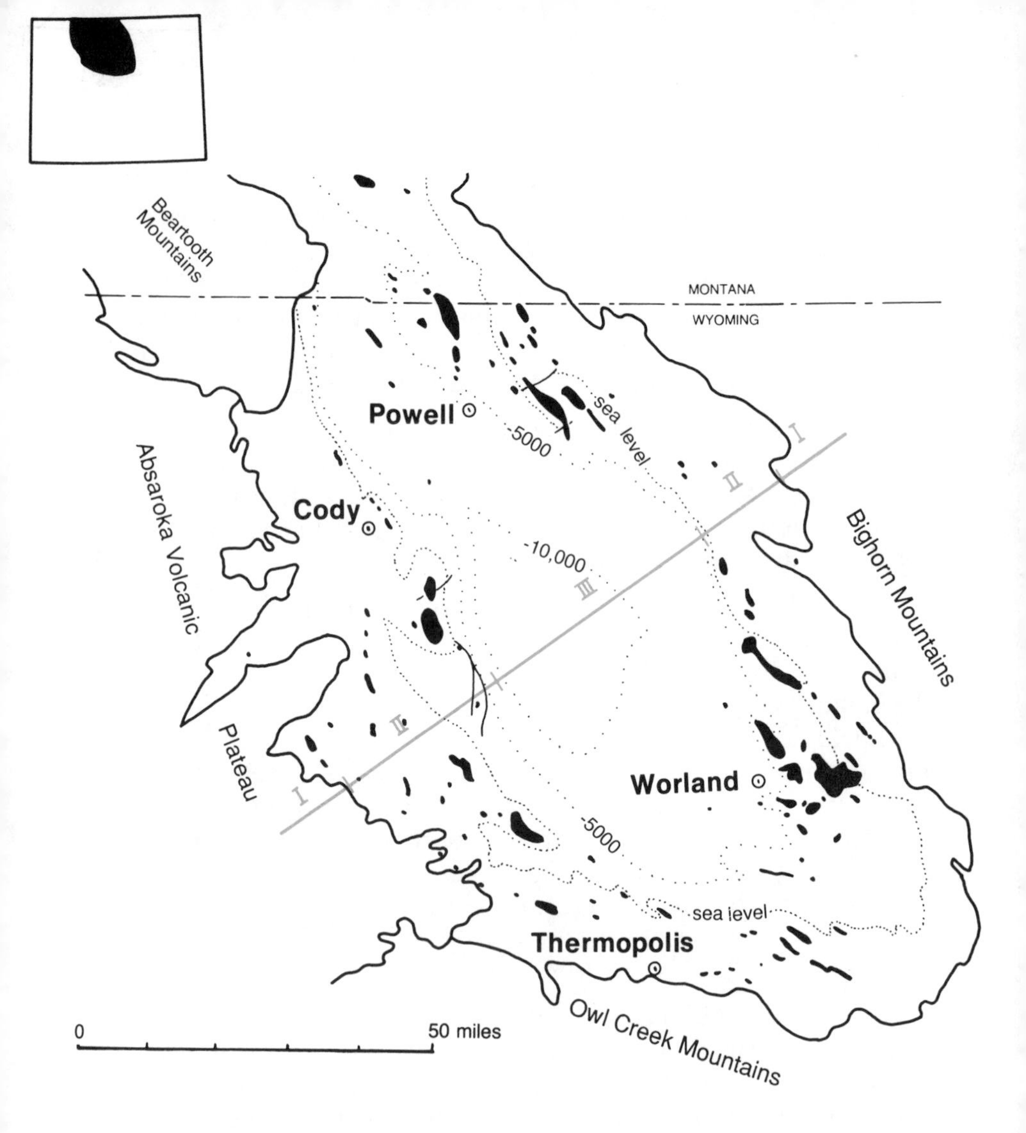

Bighorn Map showing Zone I (basin margin), Zone II (basin shoulders), and Zone III (basin trough). Oil Fields form a racetrack pattern around the shoulder. dotted lines are contours on top of the Dakota sandstone; the Dakota is a 10,000 foot deep pit east of Cody.
–Adapted from the Geological Survey of Wyoming's 1980 Oil and Gas Fields Map of Wyoming.

The third zone is the central basin trough. The west shoulder of the basin was shoved east over the central basin trough along the Oregon Basin thrust fault, which dips about 30 degrees west beneath the shoulder. The east shoulder was similarly thrust to the west over the central trough. On the surface, the central basin trough is recognized by outcrops of the

Paleocene Fort Union formation and Eocene Willwood formation. The Willwood is roughly equivalent to the Wind River, Green River, and Wasatch formations in other parts of Wyoming. These younger strata were eroded less because they occupy the lowest part of the basin.

BIGHORN BASIN OIL

Creating an oil or gas field is like baking the perfect cake; you need a few critical ingredients and the perfect balance of time and luck. A recipe for an oil field would include four fundamental ingredients. First, you need a source bed to generate the oil, such as a carbon-rich black shale. Heat and pressure over millions of years convert the carbon compounds to "hydrocarbons" of liquid oil and natural gas. Second, you need a trap to confine the oil so it doesn't spread out all over creation. Since oil is lighter than water, it typicaly floats upward along porous layers into the crest of anticlines, like those on the shoulder of the Bighorn Basin. Third, you need a porous reservoir formation within the trap that will hold the oil or gas. Oil does not fill giant underground caverns—it fills the microscopic pore spaces between grains of sand in sandstone, or tiny cavities in limestone or dolomite. Fourth, you need a seal or lid above the reservoir to prevent oil and gas from leaking to the surface. Usually shales or salt beds provide good seals because they lack porosity and permeability. If you have all four ingredients, and let them bake at 150 degrees for ten million years, you will have the perfect oily "cake!"

As mentioned previously, the shoulder of the Bighorn Basin contains numerous anticlines and synclines. A large number of these anticlines trap significant amounts of oil and are excellent oil fields. Bighorn Basin oil is typically very black and smells like rotten eggs, due to hydrogen sulfide mixed with the oil. The oil was probably generated by compaction and heating of carbon-rich, marine Paleozoic strata, like the Permian Phosphoria formation, millions of years before the Laramide orogeny. The oil may have generated in thick rock sequences in western Wyoming or eastern Idaho, then migrated up-slope to

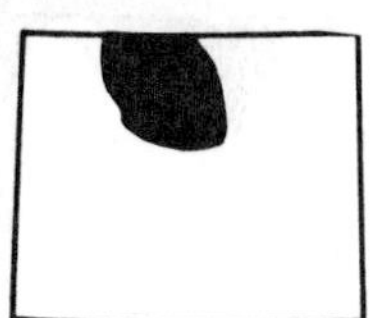

the east through the porous and permeable rock layers that now lie under the Bighorn Basin. During or after the Laramide mountain-building episode, the oil moved by "secondary migration" into anticlines and occupied porous reservoir layers like the Pennsylvanian Tensleep sandstone. Most of the "easy" oil fields were discovered long ago because they are in anticlines exposed at the surface. The modern petroleum geologist must look for new fields at deeper levels and in subtle corners of the basin.

BIGHORN BASIN BENTONITE

The Cretaceous Cody shale is the major source of bentonite, an important economic mineral. Bentonite is a special type of clay that has the unique ability to adsorb water, increasing its volume as much as 30 percent. During Cretaceous time Wyoming was covered by a shallow arm of the sea that extended north from the Gulf of Mexico to the Arctic Ocean. Ash from massive volcanic eruptions to the west drifted over Wyoming on the prevailing winds and settled into the seaway. After deposition and burial, the volcanic ash was converted to bentonite clay.

Bentonite clay is mined extensively in the Bighorn Basin. It has a wide variety of industrial uses, including its use in iron ore processing and as a "drilling mud" for the petroleum industry. Wyoming produces about 65 percent of the Nation's bentonite.

THERMOPOLIS HOT SPRINGS

Aside from Yellowstone, Thermopolis is the most famous hot springs system in Wyoming. The hot springs were used by Indians in prehistoric times, and were part of the original reservation land given to the Shoshone tribe in the 1868 Fort Bridger treaty. The land was later purchased by the Federal government and turned over to the State of Wyoming for a state park. The present park contains two commercial swimming pools, bath house, numerous springs and a variety of travertine deposits.

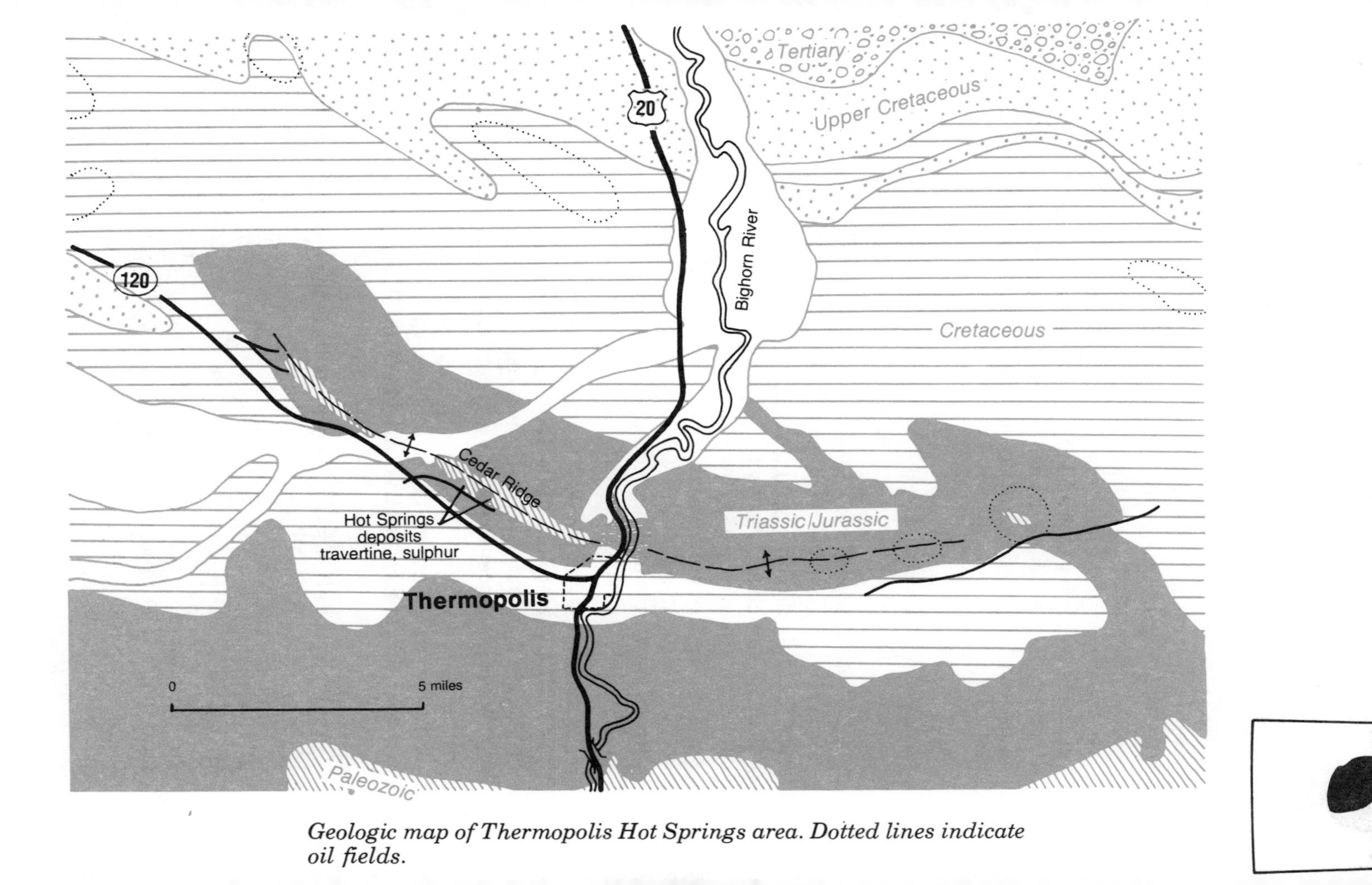

Geologic map of Thermopolis Hot Springs area. Dotted lines indicate oil fields.

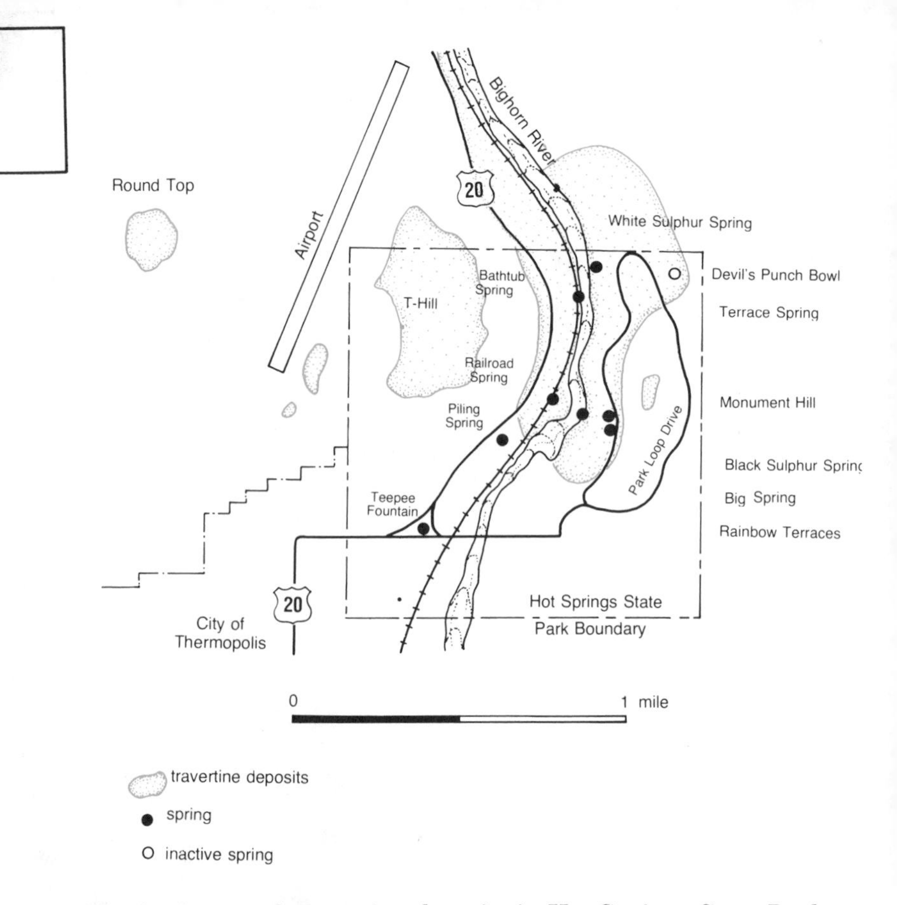

Hot Springs and travertine deposits in Hot Springs State Park, Thermopolis, Wyoming. –Adapted from Breckenridge and Hinckley (1978, p. 27)

The hot springs and town of Thermopolis are on the flanks of the Thermopolis anticline, the largest of several folds on the southern shoulder of the Bighorn Basin. Distinctive red shales of the Triassic Chugwater formation crop out in the middle of the fold. Fractured rock within the anticline provides an avenue for downward migration of ground water. The water is then heated by the naturally hotter rocks deeper in the Earth and rises back to the surface at the hot springs. Extensive deposits of travertine have been precipitated at the surface, similar to Mammoth Hot Springs in Yellowstone. While underground, the hot water dissolves calcium carbonate from limestone formations and carries it in solution to the surface, where a decrease in temperature causes precipitation of travertine. Think of Mammoth and Thermopolis hot springs as giant, natural cement factories.

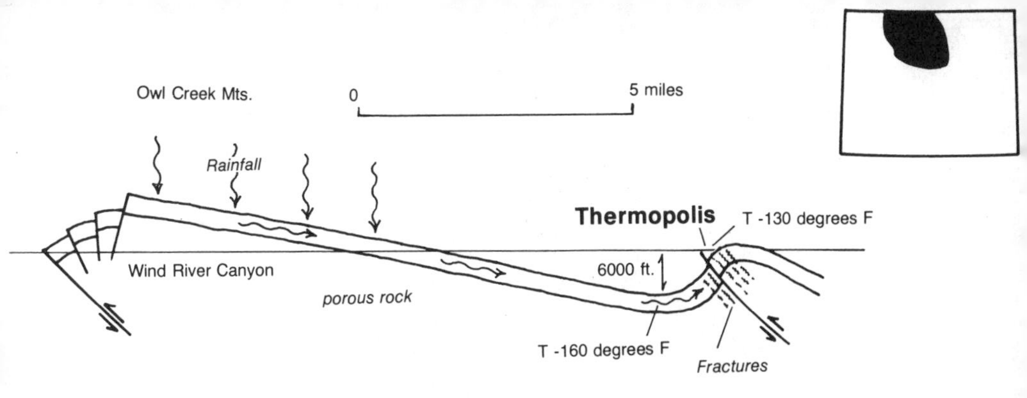

Water flow and heating system of Thermopolis Hot Springs.

Travertine terraces at Hot Springs State Park, Thermopolis.

B

BIGHORN CANYON

The Bighorn Canyon National Recreation Area straddles the Wyoming-Montana border at the north end of the Bighorn Basin. Wyoming 37, just east of Lovell, provides access. The recreation area encompasses a 71-mile-long lake created by Yellowtail Dam, built in the late 1960s, near Fort Smith, Montana. The Bighorn River has cut a spectacular canyon through the uplifted crust of the northern Bighorn-Pryor mountains. Canyon walls exposing stacks of rock layers rise above the river over 2,000 feet in places, representing more than 500 million years of Earth history. The same Paleozoic and Mesozoic formations you see in the Wind River Canyon are exposed along these canyon walls.

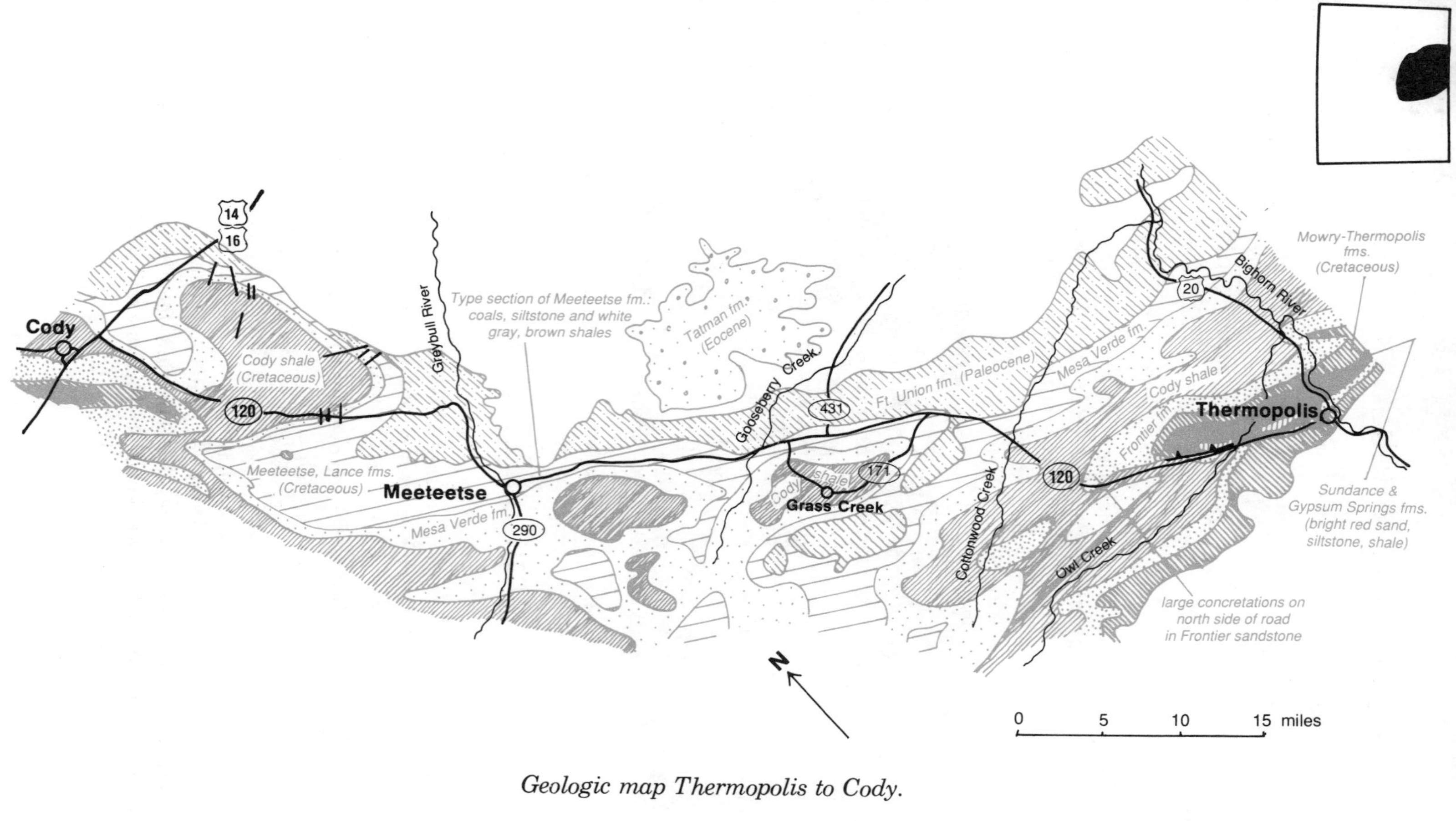

Geologic map Thermopolis to Cody.

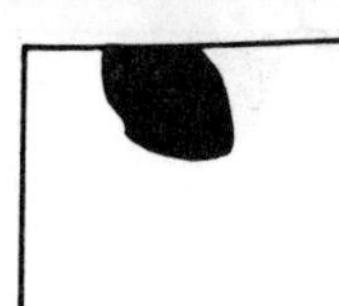

Wyoming 120
Thermopolis—Cody
84 mi./134 km.

The road between Thermopolis and Cody offers a scenic view of the southwestern Bighorn Basin. The Absaroka Mountains frame the western horizon. This is oil country—numerous oil-producing anticlines form blisters of folded strata across the landscape. Oil fields in this area, like Hamilton Dome, Grass Creek and Oregon Basin, are among Wyoming's giants. They have produced enormous quantities of oil and gas since the early 1900s.

In general, Wyoming 120 follows the eastern margin of "zone 2," or the shoulder, of the Bighorn Basin. Numerous anticlines and synclines, expressed by folded Mesozoic strata, lie west of the highway; the east-dipping Paleocene Fort Union and flat-lying Eocene Willwood formations crop out in the central basin trough to the east. Near Thermopolis the road follows the southwest limb of the Thermopolis anticline and passes steeply dipping outcrops of the Cody, Frontier, Cloverly, Morrison, Sundance, and bright-red Chugwater formations northeast of the road. The more resistant sandstone layers form pinyon-covered ridges or hogbacks.

High cliffs of brown to gray sandstone that belong to the Paleocene Fort Union formation rise east of the highway at Gooseberry Creek, four miles north of the junction with Wyoming 431. In addition to sandstone, the Fort Union formation contains gray to black shale and thin beds of coal.

Grass Creek oil field is on a large anticline west of the junction of Wyoming 431. It has produced more than 177,600,000 barrels of oil since its discovery in 1914; it is still producing.

The Eocene Willwood formation overlies the Fort Union formation farther east in the central trough of the basin. It is clearly distinguished by its brightly variegated claystones and sandstones, with some lenses of quartzite conglomerate. The Willwood and Fort Union formations are river deposits laid down after the retreat of the Cretaceous marine seaway. To the east, between Wyoming 431 and Meeteetse, the Eocene Tatman formation locally overlies the Willwood formation. It consists of drab claystone and sandstone, with some oil shale and lignite coal.

Spherical iron-stained concretions in Cretaceous Frontier formation about 14 miles northwest of Thermopolis on Highway 120.

Also between Wyoming 431 and Meeteetse, the Squaw Teats form two prominent peaks east of the highway. They are erosional remnants of the Eocene Tepee Trail formation, an olive-drab andesitic conglomerate that is 48 million years old. The andesite came from the Absaroka volcanic field to the northwest.

Meeteetse is on the Greybull River, which drains the eastern slopes of the Absaroka Mountains. The name is a Shoshone Indian word for "meeting place" or place of rest. Meeteetse was one of the first settlements in the Bighorn Basin. The tree-covered ridges around town are resistant, upper Cretaceous sandstones that dip steeply eastward into the basin. A drive up Wyoming 290 along the Wood River to Sunshine or Pitchfork offers a nice side-trip.

North from Meeteetse to the Burlington Junction, Wyoming 120 follows the floodplain of the Greybull River through outcrops of the Paleocene Fort Union formation.

Between Burlington Junction and Cody, Highway 120 crosses the western limb of the giant Oregon Basin anticline and oil field. Oregon Basin is a classic Wyoming "Sheepherder" anticline. Steeply-dipping, light-colored, massive sandstone beds within the upper Cretaceous Mesaverde formation outline the fold. Short Pinyon trees grow in cracks or open joints on these sandstones while adjacent shale layers support little vegetation. The anticline has been eroded to expose black shales of the Cretaceous Cody formation in the middle; the shales are easily eroded and form a shallow topographic hollow or basin, hence the name "Oregon Basin," although it really is an up-fold of strata. The "Oregon" part of the name has an interesting history: a cattle herd from Oregon was driven through here and an

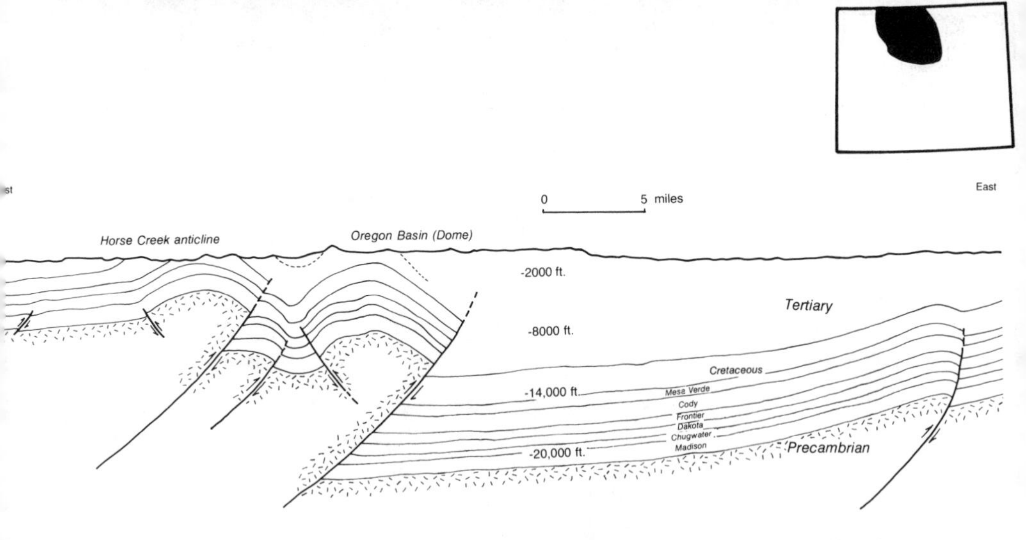

Cross section traced from a seismic section through Oregon Basin, south of Cody, Wyoming; west side of Bighorn Basin.

old, weak bull was left to die; when spring roundup came, the cowboys found the bull alive and healthy. They called him the "Oregon Bull" and his range, the "Oregon Basin."

The Oregon Basin oil field is enormous. It produces from a variety of formations ranging from Cambrian to Cretaceous in age. The field was discovered in 1912, has produced more than 360 million barrels of oil and over 174 billion cubic feet of gas, and is still going strong! Several smaller pools have been discovered around the periphery of the main field.

US 16/20
Thermopolis—Greybull
71 mi./114 ki.

This segment of highway crosses the southeastern arm of the Bighorn Basin and follows the Bighorn River as it slowly meanders north. The total distance may be divided into three segments based on the structural geology and kind of sedimentary rocks. Although these segments are presented from south to north, you can reverse their order if you are traveling the opposite direction.

The southern segment extends north from Thermopolis for about 12 miles, at which point the highway makes a sharp 45 degree bend to the northeast, just north of Kirby. This segment crosses the north-dipping limb of the Thermopolis anticline on the shoulder of the

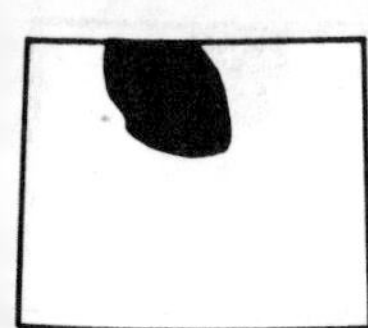

basin. From Thermopolis to the road bend, you will see, from oldest to youngest, the bright red Triassic Chugwater formation, Jurassic Sundance and Morrison formations, and the Cretaceous Cloverly, Frontier, Cody, Mesaverde, Meeteetse, and Lance formations. Sandstone layers within these formations, especially the Mesaverde, form hogback ridges that have their steeper sides facing south.

The central road segment extends from just north of Kirby to Manderson, and crosses the central basin trough. The Bighorn River has cut a wide floodplain into the Paleocene Fort Union formation and the Eocene Willwood formation. These units are described in the Thermopolis-Cody road section. Quaternary gravels cap the flat-topped terraces above the floodplain and east of Worland. Just east of Worland is an enormous complex of oil fields that produce from deeply buried Permian and Pennsylvanian strata.

The third road segment extends from Manderson to Greybull. The highway straddles the western margin of the basin's shoulder and cuts into upper Cretaceous strata near Greybull. Remember that the east shoulder of the basin was shoved westward over the central basin trough. This thrust fault lies deep beneath the surface and is not exposed; however, its general trend is north-south, about parallel to the road between Manderson and Greybull. The large oil field at Manderson produces from Pennsylvanian, Permian, and lower Cretaceous strata.

Greybull is on the western edge of the basin's east shoulder. Nearly flat-lying outcrops of Paleocene and Eocene strata extend across the basin to the west, whereas folded Mesozoic strata on the shoulder extend east to the base of the Bighorn Mountains. Sheep Mountain anticline, one of the best exposed anticlines in the Rocky Mountains, lies just north of town. The bright red Triassic Chugwater formation defines the south end of the anticline. Greybull was founded in 1909 as an agricultural center and was named for the river, which the Indians named after an albino buffalo that roamed the area. Albino buffalo were sacred to the Indians.

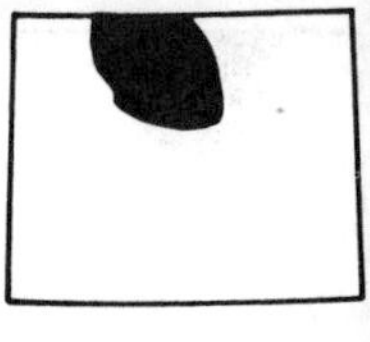

CENTRAL AND NORTHERN BIGHORN BASIN

US 16/20
Greybull—Cody
53 mi./85 km.

The highway between Greybull and Cody offers an almost complete cross section of the Bighorn Basin. As you drive across the basin, you can sense that it is a large syncline, or downwarp, between the Bighorn and Beartooth-Absaroka mountains. Both Greybull and Cody are on the encircling shoulder of the basin, which we called "zone 2" in the introduction.

Greybull is on alluvial sediments of the Bighorn River, which overlie black shales of the upper Cretaceous Cody formation. Resistant sandstone beds of the Cretaceous Frontier formation crop out in the ridges immediately east of town. West of Greybull, the highway crosses narrow outcrops of west-dipping, uppermost Cretaceous strata. An old erosion surface truncated them, and then the strata were buried beneath the Paleocene Fort Union formation. This unconformity is well-exposed south of the highway about 3 miles west of the Bighorn County Fairgrounds. It tells us that the upper Cretaceous rocks were tilted west into the basin during the early part of the Laramide orogeny, then eroded and truncated by streams before Fort Union sediments were deposited. Farther west, the brown and gray sandstones of the Fort Union formation are overlain by vari-colored claystones of the Eocene Willwood formation, which crop out from the central basin trough to the east flank of Oregon Basin anticline. Basin subsidence and mountain uplift that continued into early Eocene time locally tilted the Willwood formation along the margins of the basin.

US 16/20 skirts the north end of the Oregon Basin anticline along the west end of this traverse. Oregon Basin is discussed in the section from Thermopolis to Cody.

Cody

Cody lies at the east base of Rattlesnake Mountain along the Shoshone River, within the western shoulder of the basin. The town is

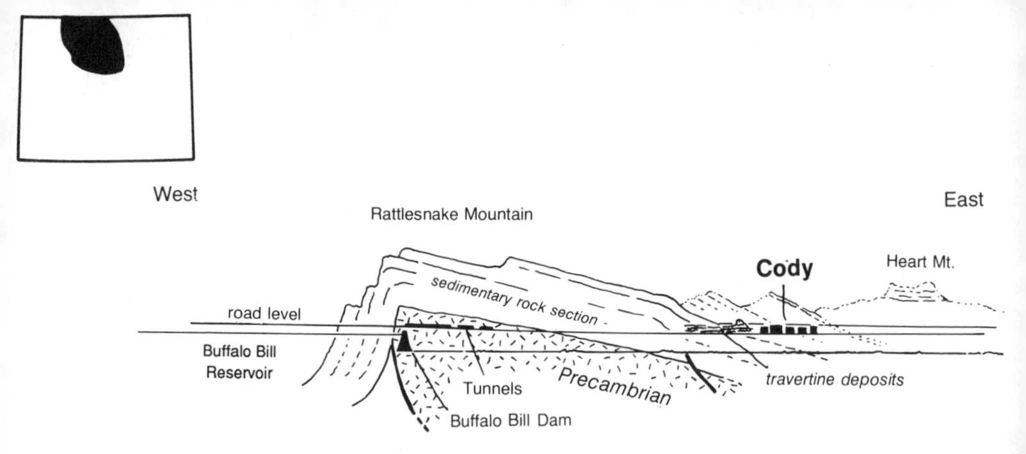

Cross section sketch of Rattlesnake Mountain and Shoshone Canyon west of Cody, Wyoming.

named for the famous western entrepreneur, William F. "Buffalo Bill" Cody. It is the door to eastern Yellowstone and home of the world famous Buffalo Bill Historical Center, a showcase of western history, art, and Indian culture. Cody is a lively western town that has managed to retain some of the flavor of the Old West. The old Irma Hotel, built in 1903 by W.F. Cody and named for his daughter, still serves meals and offers lodging.

Geologically, the town is on Mesozoic strata that dip east off the backside of Rattlesnake Mountain. These strata were planed off by erosion, then covered by much younger Quaternary gravels to form the "Cody terraces." The angular unconformity between the underlying Mesozoic sandstones and shales, and overlying terrace gravels is well exposed at two spots: on the north side of town where Wyoming 120 crosses the Shoshone River, and just west of town along U.S. 14/20 where the river gorge can be seen to the north. These Quaternary terraces are composed of gray to black cobbles of Eocene volcanic

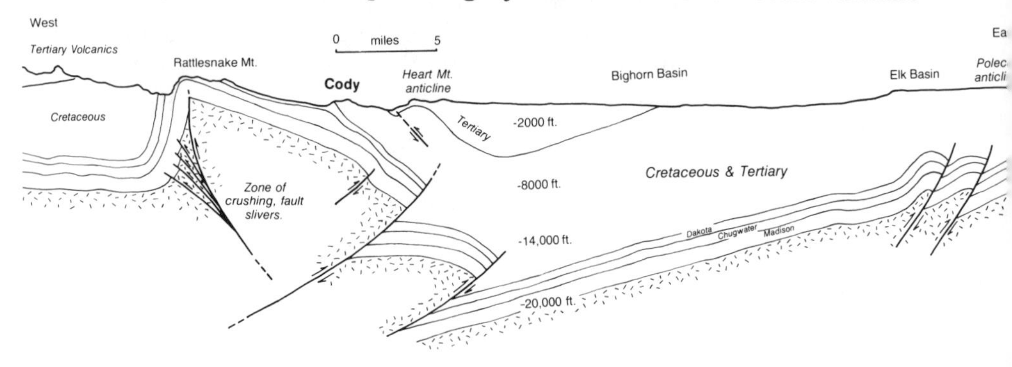

Cross section interpreted from a seismic section near Cody, Wyoming.

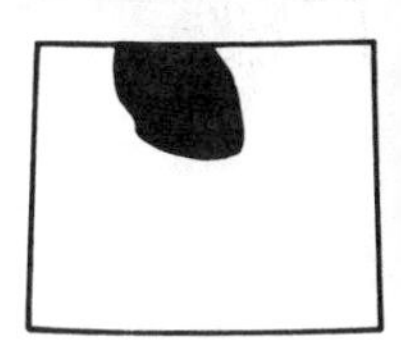

Angular unconformity formed by Quaternary gravels overlying tilted Mesozoic shales and sandstones at Cody. Heart Mountain in the distance.

rocks that were carried east from the Absaroka Mountains by the Shoshone River and deposited before the river cut its present gorge.

On a hot summer day, the sulfurous odor of rotten eggs fills the air along the Shoshone River west of Cody. The smell comes from thermal springs along the river banks. Mountain man John Colter, the first white man in the Bighorn Basin, brought back stories of the "Stinking Water River," later renamed the Shoshone. The springs issue from large fractures in the Permian Phosphoria formation, which is mostly limestone in this area. Geologists propose that water flows eastward underground from the high Absaroka Mountains under artesian pressure, is heated at depth, then forced along faults and fractures associated with Rattlesnake Mountain. The springs rise through the Phosphoria formation at this point because Mesozoic formations immediately to the east contain an abundance of impermeable shales that block the water. A similar spring that existed on the southwest side of Rattlesnake Mountain was inundated in 1907 when Buffalo Bill Reservoir was filled. Layers of travertine, formed from old springs, are seen west of Cody.

Travertine deposits unconformable over Mesozoic section at Pullout, west side of Cody on Highway 14-16-20.

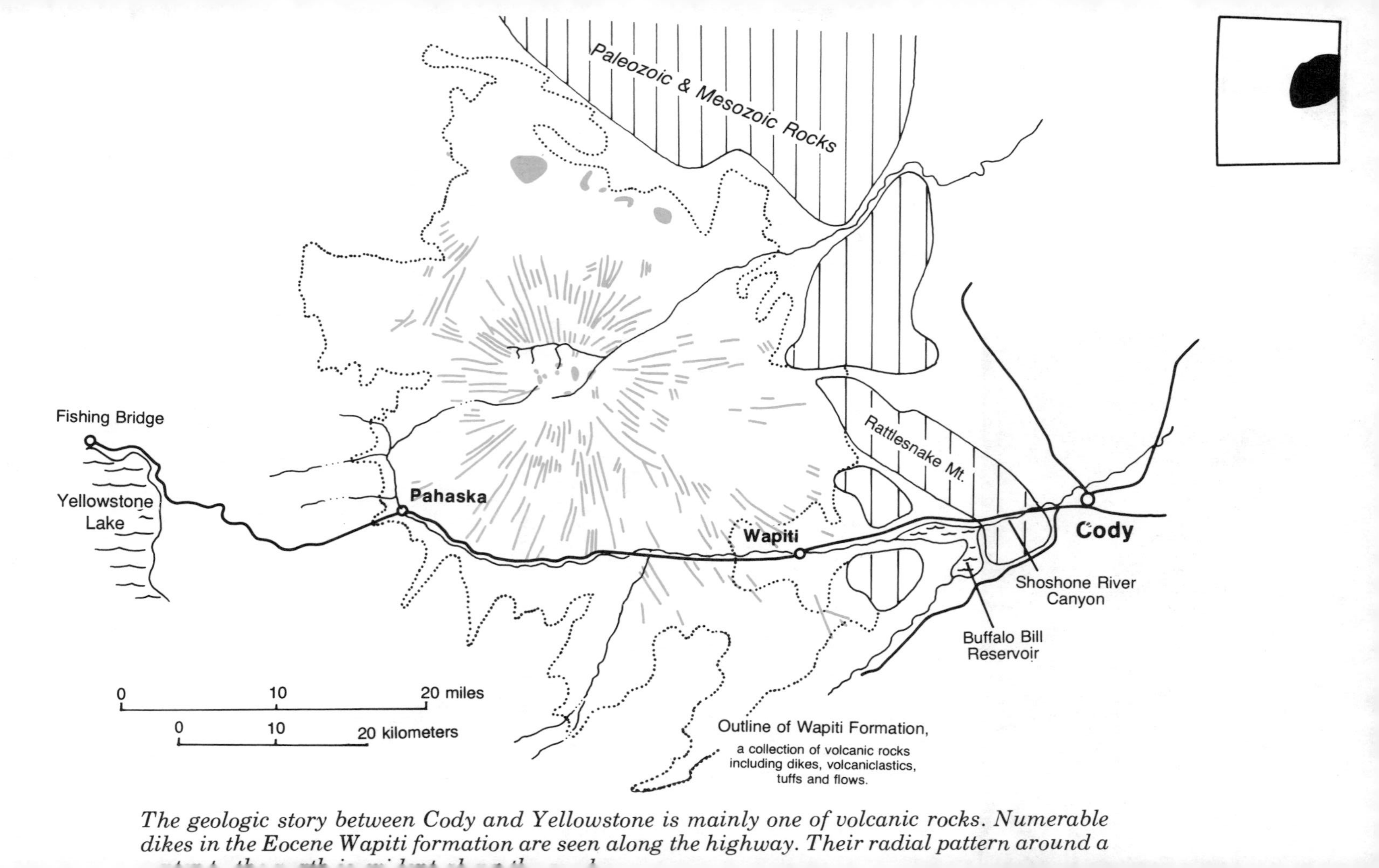

The geologic story between Cody and Yellowstone is mainly one of volcanic rocks. Numerable dikes in the Eocene Wapiti formation are seen along the highway. Their radial pattern around a

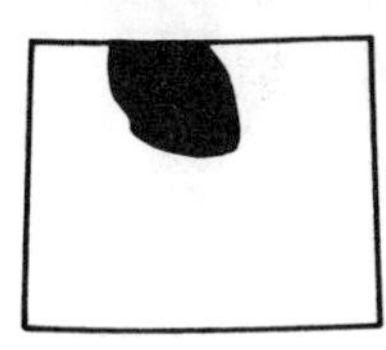

Rattlesnake Mountain west of Cody on Highway 14-16-20.

US 14/20
Cody—East Entrance of Yellowstone Park
52 mi./83 km.

This highway follows the beautiful valley of the Shoshone River from the semi-arid landscape of Cody to the high, wet forest of the Absaroka Mountains. The Shoshone National Forest was the first National Forest in the U.S., established by President Harrison on March 30, 1891. Also, the first Forest Service Ranger Station was built along the Shoshone River in 1903. Geologically, the Eocene volcanic rocks that form the massive Absaroka Mountains are seen for the entire distance, except at the east end where the road cuts through Rattlesnake Mountain.

Rattlesnake Mountain forms a sharp ridge that trends northwest. It is an anticline formed when the overlying Paleozoic and Mesozoic sedimentary rocks were arched by rise of a block of Precambrian metamorphic and igneous rocks. The anticline is asymmetric to the west, meaning its western limb is nearly vertical, whereas the east limb dips gently towards Cody. As you drive through Shoshone River Canyon west of Cody, the cliffs high on the canyon wall expose hard strata of Paleozoic age: the Pennsylvanian Tensleep sandstone and Mississippian Madison limestone. Lower on the canyon wall, a massive white cliff is the Ordovician Bighorn dolomite. The highway tunnels were carved through Precambrian metamorphic rocks in the core of the mountain. On the west side, you emerge from a tunnel at the Buffalo Bill Dam and again encounter strata of Paleozoic age, only these beds are much more steeply inclined and dip to the west. You have driven through the core of a large anticline!

Some of the early geologists who studied Rattlesnake Mountain

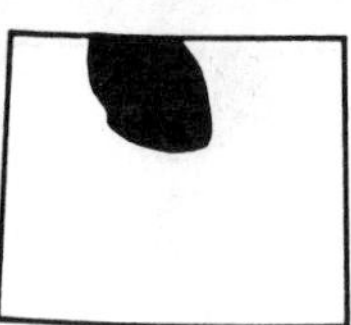

suggested that the west flank was bounded by a vertical normal fault. Other geologists believed that this anticline, and others like it in Wyoming, were bounded by thrust faults that raised the Precambrian "basement." Geologists tend to be argumentative, and these interpretations were hotly debated for several years. The authors believe, as most workers now do, that Rattlesnake Mountain is bounded by a thrust fault along its steep west flank that dips east beneath the uplift. Similar interpretations have proven true elsewhere in Wyoming, where oil wells have been drilled through Precambrian rocks into Paleozoic and Mesozoic strata below a thrust fault. The implications are very important to petroleum geologists looking for undiscovered sedimentary traps that contain oil or gas!

Buffalo Bill Dam, the first concrete arch dam in the world, was constructed in 1910 for hydroelectric power, downstream irrigation, and recreation. The dam site is in the narrowest part of the canyon where the Shoshone River bit deep into very hard Precambrian igneous and metamorphic rocks. The reservoir extends west over a broad valley underlain by soft shales of the Cretaceous Cody formation, the same unit that lies beneath Cody on the east side of Rattlesnake Mountain anticline. The North and South forks of the Shoshone River converge at Buffalo Bill Reservoir to give the lake its distinctive heart shape. Sheep Mountain separates these two drainages. It is the darkly forested plateau south of the highway and west of Rattlesnake Mountain. The upper part of Sheep Mountain contains Paleozoic strata that slid over younger Eocene rocks, which form the lower slopes; this is part of the "Heart Mountain Detachment" story, one of the most puzzling geological features in northwestern Wyoming. We discuss it in a following section.

West of Buffalo Bill Reservoir, the lower slopes near the highway are outcrops of vari-colored claystone and sandstone of the Eocene Willwood formation, overlain by young landslide deposits of mixed rock and dirt. The higher slopes and surrounding mountains are in the Eocene Wapiti formation, reddish-brown deposits of brecciated volcanic rocks, mostly andesite. These breccias were deposited as soupy mudflows within the Absaroka volcanic field around 40 million years ago. The 1980 eruption of Mount St. Helens in the Cascade volcanic chain of Washington gave geologists a good model to use in interpreting these ancient volcanic deposits. During and after the eruption, spectacular mudflows of fragmented volcanic rock, ash and water flowed down the valleys away from Mount St. Helens. Close

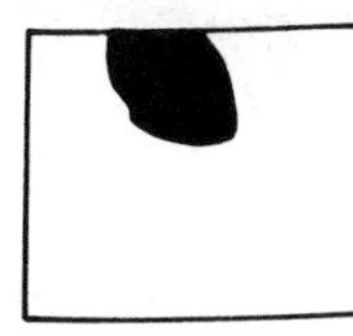

Dikes in the Eocene Wapiti volcanics that form "Chinese walls" about 20 miles west of Cody on Highway 14-16-20.

your eyes, if you're not the driver, and imagine this area 50 million years ago—volcanos similar to Mount St. Helens erupt one after the other with thick, muddy mixtures of rock and ash debris flowing away. Several pull-offs along the upper part of the Shoshone valley let you stop to inspect these ancient mudflow breccias—notice the different colors of the volcanic fragments; the color is caused by the chemical composition of the fragment plus later alteration and weathering. In some areas, the breccias have been eroded to form cone-shaped pinnacles, or "hoodoos." Also, look for volcanic dikes intruded into cracks after the mudflow breccias were deposited. These dikes look like miniature versions of the "Great Wall of China," because they form resistant, vertical walls of volcanic rock on the sides of hills.

A small, river-carved cave is near the mouth of Blackwater Creek, on the north side of the road. The mummified body of a man who died

Chimney Rock formed by erosion of Eocene Wapiti volcanics along a vertical joint about 30 miles west of Cody on Highway 14-16-20.

1,300 years ago was found in this cave. Archeologists believe the cave was used off and on by Indians for 9,000 years.

The road log for the east entrance of Yellowstone to Fishing Bridge is in the chapter on Yellowstone National Park.

Wyoming 120
Red Lodge, Montana—Cody
62 mi./99 km.

From Red Lodge to Belfry, Montana, Montana 308 descends through coal-bearing strata of the Paleocene Fort Union formation, which consists of stream-deposited, brown and gray sandstone and gray shale, and swamp-deposited coal beds. Thin coal beds are exposed in stream cuts south of the highway. These beds gently dip west toward the Beartooth Mountains, which were raised and shoved over the sedimentary layers along a thrust fault.

Washoe was the site of one of the worst underground coal mine disasters in history. On February 27, 1943, a violent explosion occurred in the Smith Mine. The blast, along with lethal methane gas, killed 74 miners. Searchers found this farewell message:

> "Walter and Johnny. Goodbye wives and daughters. We died an easy death. Love from us both. Be good."

From Belfry south to the Wyoming State line, Montana 72 (Wyoming 120) follows the Clarks Fork River upstream as it cuts through buff sandstones and gray-green shales of the Fort Union formation on the central floor of the northern Bighorn Basin. These sediments appear in roadcuts north of the state line.

Wyoming 120 cuts through the Badger Basin oil field just south of the state line. It has produced almost 3,000,000 barrels of oil from the Cretaceous Frontier and Dakota formations since its discovery in 1930.

Between the state line and the intersection with Wyoming 292 to Clark, Highway 120 gradually climbs through the sedimentary rock section to the vari-colored red and green claystones of the Eocene Willwood formation; the contact between Fort Union and Willwood lies near the intersection of Wyoming 120 and Wyoming 292 to Clark. Aside from recent Quaternary deposits, the Willwood, about 50 mil-

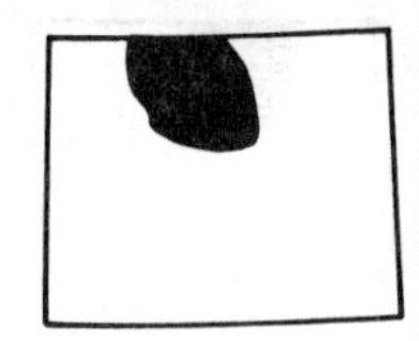

Range-front monocline of Paleozoic and Mesozoic strata that form the front of the Beartooth Mountains Clarks Fork Canyon. This section is easily visible from Highway 120 north of Cody.

lion years old, is the youngest sedimentary formation in this part of the basin; it is partially preserved from erosion because it lies in the central trough of the Bighorn Basin. Older units along the basin margins are truncated or completely eroded away. The Willwood contains abundant fossil remains, mostly teeth, of early mammals and other vertebrates.

To the west is the mouth of the Clarks Fork Canyon. Wyoming 292 offers a short side-trip with a beautiful ending. The Clarks Fork of the Yellowstone River has cut a deep, narrow canyon in Precambrian metamorphic rocks on the southeast corner of the Beartooth Mountains. Beds of Paleozoic limestone were folded over the edge of the

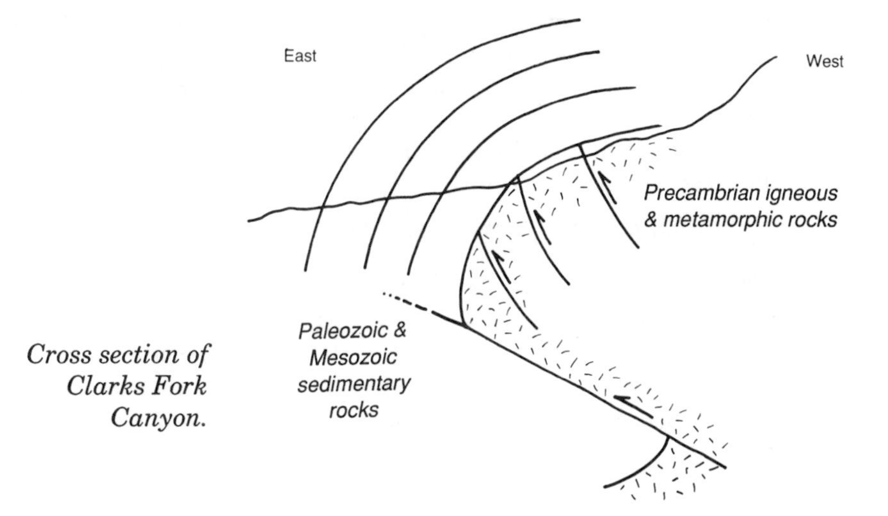

Cross section of Clarks Fork Canyon.

Beartooth uplift, then eroded to form spectacular, vertical, gray cliffs, which resemble the "flying buttresses" of a Gothic cathedral.

Back on Wyoming 120 to Cody, Heart Mountain is the prominent peak on the horizon to the south. It is composed of Paleozoic strata that slid downhill from the Beartooth Mountains over the Eocene Willwood formation. A few miles north of Heart Mountain, look south at the Willwood strata beneath Heart Mountain; you will see a broad, gentle, down-folded syncline that contains the axis, or structurally lowest part, of the Bighorn Basin.

Wyoming 120 follows the western shoulder of the Bighorn Basin from the Sunlight Basin road, Wyoming 296, to Cody. Remember from the introduction to the basin that the entire western shoulder was shoved eastward over the basin by the Oregon Basin thrust fault. For several miles south of the Sunlight Basin road, Wyoming 120 follows prominent, massive, shoreline and near-shore marine sandstone beds of the upper Cretaceous Mesaverde formation east of the highway. The Swiss-cheese appearance of some of the thick beds is due to chemical weathering. The Mesaverde sandstone is overlain to the east by uppermost Cretaceous strata and non-marine Paleocene and Eocene sandstones and claystones, which are finally capped by the slide mass of Heart Mountain.

Just west of Heart Mountain, the Mesaverde formation is folded into a small anticline and syncline, forming Heart Mountain oil field. This field has produced approximately 44,000 barrels of oil and 46 million cubic feet of gas from the Cretaceous Frontier formation.

About five miles north of Cody, the road cuts through red and green shales of the Jurassic Morrison formation and steeply inclined gray sandstones of the lower Cretaceous Cloverly formation. These units were folded on the flanks of two small anticlines in the area, which typify the shoulder of the Bighorn Basin.

Heart Mountain Detachment Fault

The Heart Mountain detachment fault is, without doubt, one of the most puzzling geologic features in North America. No one has yet proposed an "airtight" explanation of how it formed! But first, what is it?

The Heart Mountain detachment fault is a low-angle fault on the

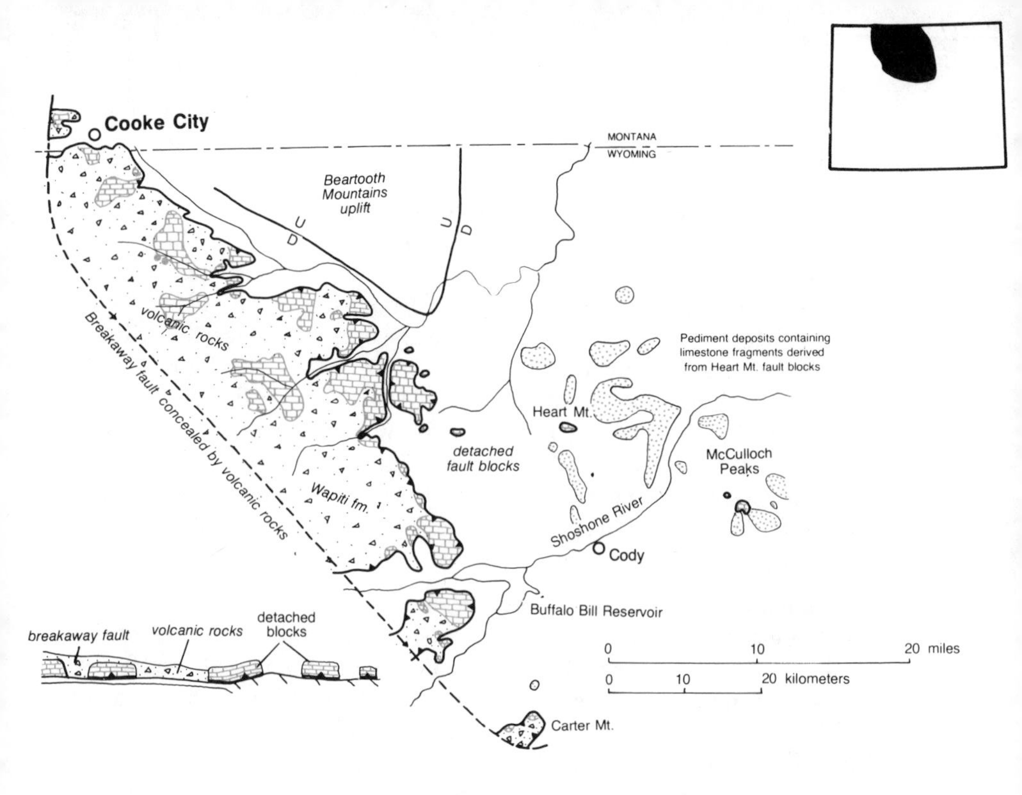

Map and cross section of Heart Mountain detachment, where immense blocks of Paleozoic carbonates slid along a surface for many miles.
–Adapted from Pierce (1987, p. 149)

south margin of the Beartooth Mountains that extends over the northwest corner of the Bighorn Basin. The fault plane is at least 100 miles long. It extends from south of Cooke City, Montana, to the Carter Mountains south of Cody, and 10 miles east of Cody to the McCulloch Peaks. It is named after Heart Mountain 15 miles north of Cody. In essence, the Heart Mountain detachment fault is a giant sliding surface. Blocks of Paleozoic strata moved across this surface for tens of miles. Heart Mountain, the McCulloch Peaks, Carter Mountain, Sheep Mountain and dozens of other blocks in the Cody vicinity moved great distances over the detachment fault to reach their present positions.

The detachment fault may be subdivided into four segments. At the northwest end, the first segment is a vertical "break-away" scarp, similar to the break-away zone at the top of a snow avalanche or a slump. The second segment is below the break-away scarp and is called the "bedding-plane fault," because it occurs parallel to the sedimentary beds. Oddly, the bedding-plane fault occurs exactly 8 feet above the base of the Ordovician Bighorn dolomite along an ordinary looking bedding contact. The third segment is called the "transgressive fault" because the slide blocks moved over the eroded and truncated ends of younger strata. The transgressive fault is at Dead Indian Hill a few miles north of Cody; the north side of Dead Indian Hill contains many jumbled blocks, some the size of buildings, that never made it over the transgressive fault. The fourth fault segment is the former land surface of the Bighorn Basin. The detached blocks of limestone and dolomite moved over the floor of the basin and littered hundreds of square miles. If you drive 15 miles north of Cody on Wyoming 120 and take Wyoming 296 over Dead Indian Hill to Cooke City, you traverse all four segments of the Heart Mountain detachment fault.

When did all this sliding take place and what caused it? These are hard questions to answer because erosion has destroyed some of the evidence, and younger rocks have covered more. Geologists generally agree that most of the sliding occurred in Eocene time, about 45 to 50 million years ago. As the blocks detached and slid towards the basin, Eocene volcanic rocks filled the spaces between them before the fault plane could be eroded. This shows that the sliding occurred before the Eocene sediments were deposited. Also, Heart Mountain and other slide blocks rest on the Eocene Willwood formation in the basin, so it had to be deposited before sliding took place. These relationships "bracket" the sliding during the Eocene.

However, the million dollar question remains—what caused the sliding? How could the blocks move over the transgressive segment of the detachment fault and stay in one piece as they moved? Some geologists propose that since the detachment fault extends from the mountain to the basin, the blocks were pulled downhill by the force of gravity, like a giant, slow moving landslide. This explanation is not entirely satisfactory because in landslides rocks become mixed, jumbled and broken, and that is not the case here. Also, tremendous forces of friction on the fault must be overcome for sliding to take place. Other theories propose fluid pressure along the fault buoyed the overlying slide blocks and allowed movement, or volcanic gases from the nearby Absaroka volcanic field were injected along the fault plane and reduced friction, or earthquakes literally shook the blocks down the mountain onto the basin floor. Every geologist that visits Heart Mountain has an opinion, but so far, the Heart Mountain detachment defies a completely satisfactory explanation.

Volcanic "plug" or neck southwest of Cody on Highway 291.

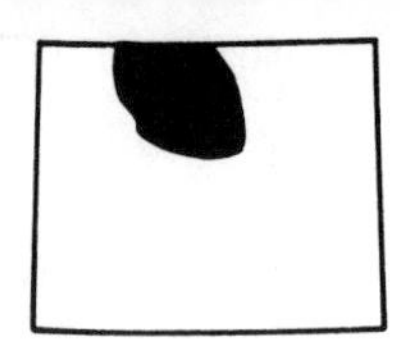

Alternate 14
Lovell—Cody
48 mi./77 km.

Lovell is on the eastern folded shoulder of the basin and Cody is on the western folded shoulder. Between the two, Alternate 14 follows the general course of the Shoshone River as it meanders across the basin to join the Bighorn River. The central part of the northern Bighorn Basin is floored by vari-colored red claystones of the Eocene Willwood formation.

The town of Powell, halfway between Lovell and Cody, was named after John Wesley Powell, explorer of the Colorado River drainage and early director of the U.S. Geological Survey.

The northeastern shoulder of the basin has been extensively folded into a series of oil-producing anticlines in the area of Byron, Lovell, and Deaver. These folds are seen at the surface as variably inclined Cretaceous sandstone layers.

The Byron oil field has produced over 116,500,000 barrels of oil and 12,800,000 cubic feet of gas since its discovery in 1918. This is a very oily corner of the Bighorn Basin!

VI

Overthrust Belt

INTRODUCTION

Westernmost Wyoming, south of the Tetons and north of the Uinta Mountains, is an area of sharp contrasts. The landscape varies from the snow-covered peaks of the Salt River, Wyoming, and Hoback ranges, to the peaceful, fertile Star Valley, to the sage-covered high desert around Kemmerer and Evanston. This region has some of the most spectacular structural geology in the state, as well as some of the largest oil and gas fields. For the geologist, this part of Wyoming is a paradise of structural intrigue and stratigraphic variation; nothing seems constant or uniform.

The overthrust belt, or fold and thrust belt, is a series of large overthrust sheets of rock that overlap one another like shingles on a roof. These thrust sheets and associated thrust faults are the result of compression and contraction of the crust during a period of mountain building called the Sevier orogeny, that continued from approximately 150 to 55 million years ago. The name Sevier comes from the Sevier Desert in central Utah where this style of deformation also exists. The Wyoming overthrust belt and Sevier Desert of Utah are but two small segments of an enormous, sinuous trend of thrust faults and folds that forms the backbone of the continent along the 5,000 miles from Alaska to Mexico.

The overlapping thrust sheets of the overthrust belt lie on westward sloping thrust faults. Thrust faults are generally the product of compression acting horizontally through the crust,

Location of Wyoming along the Overthrust and Foreland Belts of North America.

squeezing it shorter. The individual thrust faults merge at depth into a regional zone of detachment, called a decollement zone, typically in a weak layer of rock, like shale; all the layers above this detachment zone were shortened and shingled as the crust was slowly compressed through long periods of geological time. This type of thrusting is sometimes called thin-skinned because the faults are relatively shallow and flat, and do not cut into Precambrian basement rocks. Many individual thrust faults are shaped like stair steps because they had to cut through alternating hard and soft layers of rock, which forced the overlying strata to fold over the steps. These folds, or ramp anticlines, are important to petroleum geologists because they can serve as traps for oil and gas, as in the southern part of the Wyoming overthrust belt near Evanston. Some ramp anticlines are large enough to form entire mountain ranges, such as the Salt River range east of Afton.

Many major thrust faults apparently formed in an east-to-

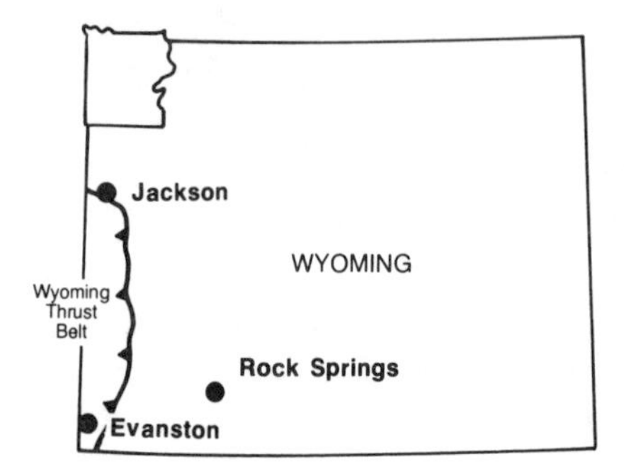

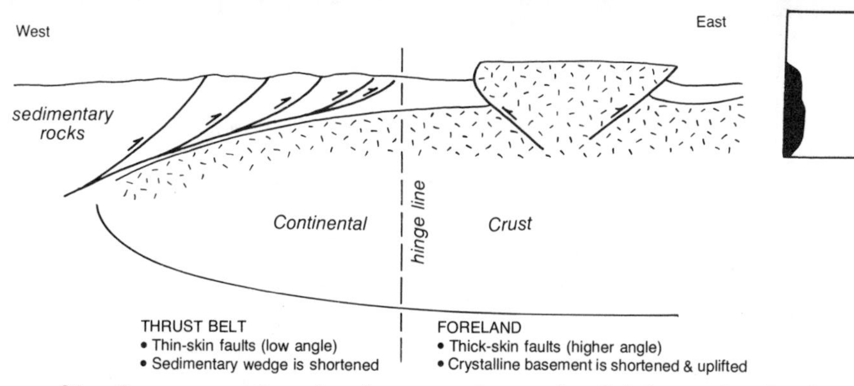

Simple cross section sketch comparing style of deformation in the Thrust Belt and the Foreland.

west sequence during the late Mesozoic and early Cenozoic. Although still in some doubt, the following sequence of thrusting is now widely accepted:

1st to move—Paris-Willard thrust fault ......Early Cretaceous
2nd to move—Crawford-Meade thrust fault...Mid-Cretaceous
3rd to move—Absaroka thrust fault..............Late Cretaceous
4th to move—Darby thrust fault..............................Paleocene
5th to move—Prospect-Jackson thrust...............Early Eocene

Much of the valley topography of western Wyoming was created by recent normal faults that came long after the thrust faulting. Normal faults generally form by extension of the crust by tensional, pull-apart stresses. This tension is occurring today throughout the Great Basin Province of the western United States, creating a topography of alternating uplifted ranges and dropped valleys, referred to as basin and range structure. Western Wyoming straddles the eastern margin of the Great Basin and lies within a zone of earthquake activity known to geologists as the intermountain seismic belt. Many of the young normal faults followed older thrust faults where they stepped up or ramped at steep angles towards the surface.

Cross section sketch showing how the Thrust Belt builds eastward, creates a series of thrust sheets, and shortens the crust in the process. A is earliest, C is latest.

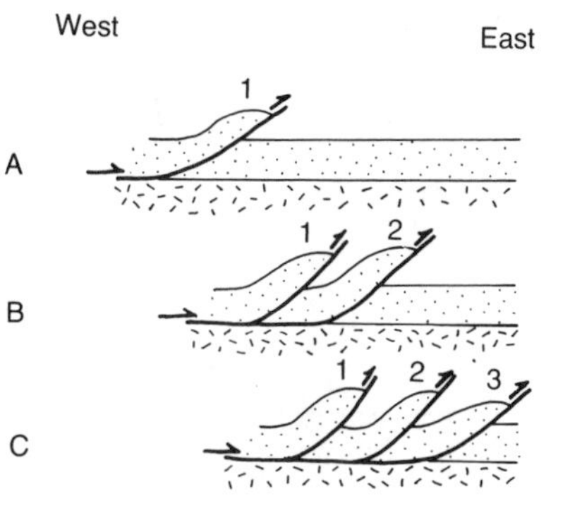

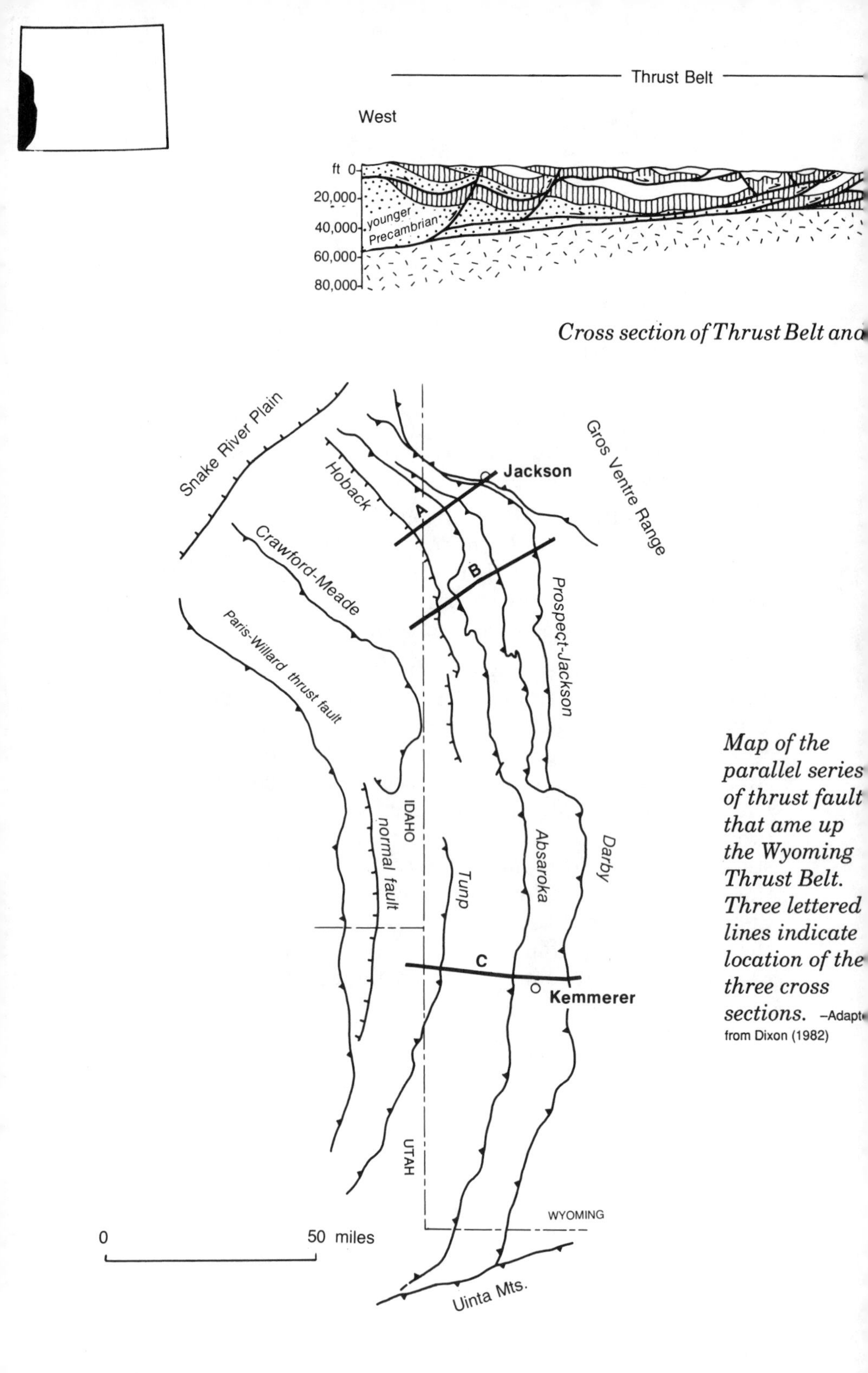

Cross section of Thrust Belt and

Map of the parallel series of thrust fault that ame up the Wyoming Thrust Belt. Three lettered lines indicate location of the three cross sections. –Adapt from Dixon (1982)

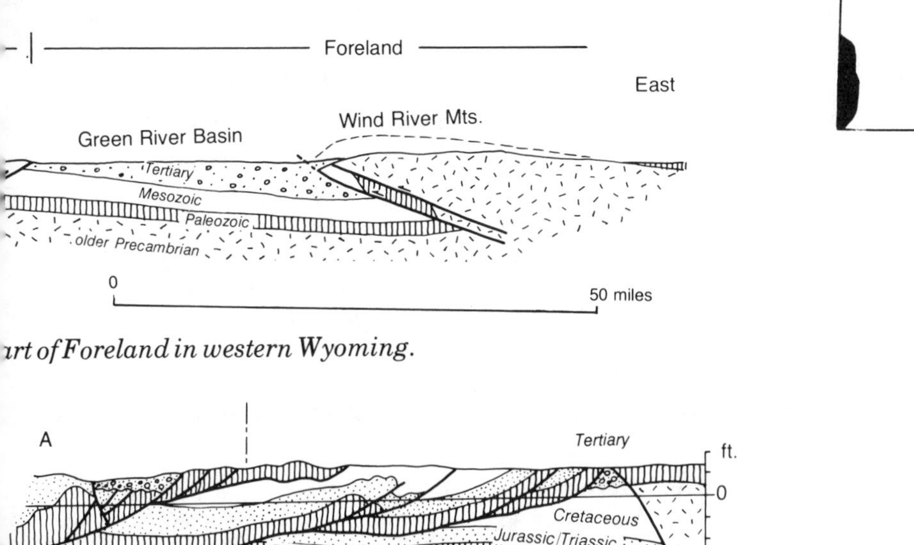

ırt of Foreland in western Wyoming.

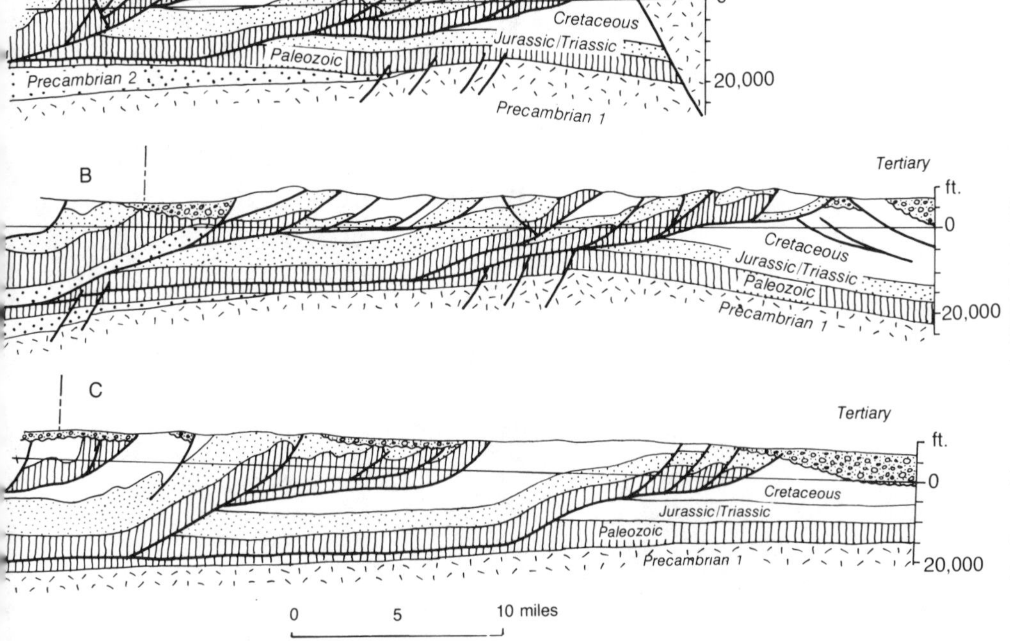

Cross sections, from north to south, across the Wyoming Thrust Belt.
–Adapted from Dixon (1982)

As a normal fault moves, one side drops to form a valley while the other rises to form a mountain range. The Hoback River Valley and Star Valley are good examples.

US 189, US 30
Interstate 80—Kemmerer
35 mi./56 km. 41 mi./66 km.

There are two ways to go to Kemmerer from Interstate 80. US 189 follows the west side of Oyster Ridge, formed by the Cretaceous Frontier formation. US 30 from Little America follows the Hams Fork River through outcrops of the Eocene Bridger formation, a section of olive-drab and white tuffaceous sandstone, and the light tan, laminated marlstone and sandstone of the Green River formation.

The giant Kemmerer Coal Company mine is south of Diamondville. Coal seams are exposed in the cut-away cliffs above the plant.

US 89/30
Kemmerer—Cokeville—Afton—Alpine Junction

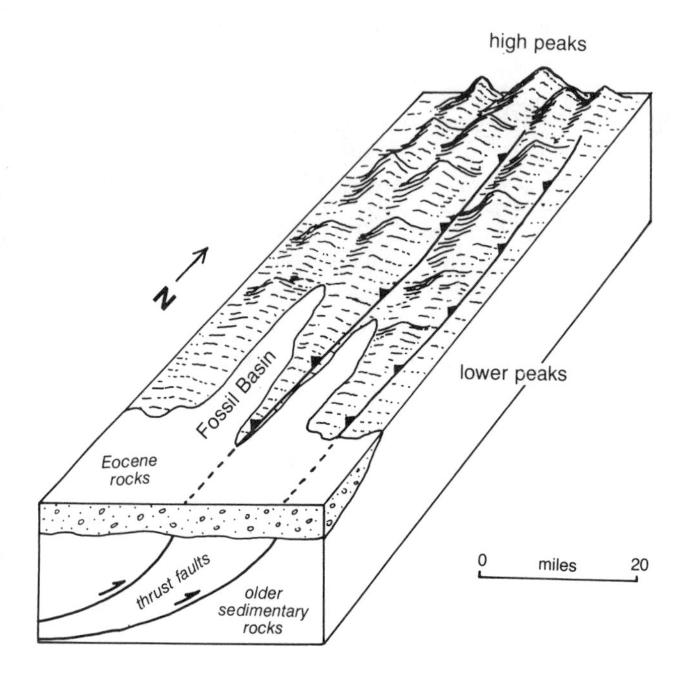

Diagram of the setting of the Fossil Basin. Note how the Eocene lake and stream sediments fill in the low places. Note also how the topography of the thrust belt mountains plunges to the south, producing lower and lower relief.

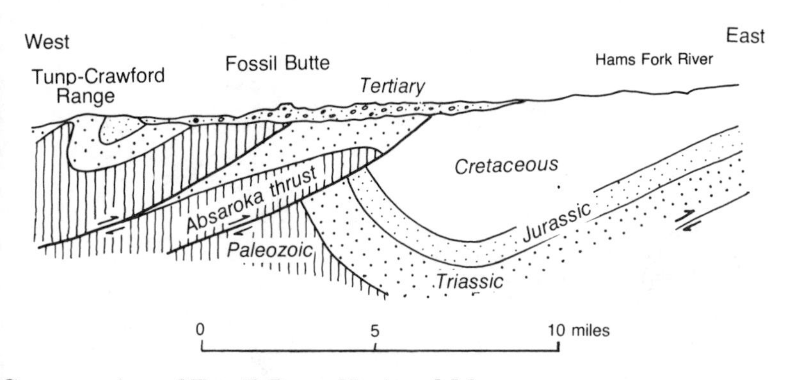

Cross section of Fossil Butte National Monument. –Adapted from McGrew and Casilliano (p. 18)

KEMMERER—SAGE JUNCTION
Fossil Butte National Monument
26 mi./42 km.

The highway between Kemmerer and Sage Junction traverses Oyster Ridge on the east and the Tunp ovethrust slab on the west, with Fossil Basin in the middle.

Hills east of Fossil Basin are west-dipping strata in the slab that moved on the Darby, or Hogsback overthrust, the easternmost overthrust in southwestern Wyoming. Three main formations exist in these hills, all Cretaceous in age; 1) on the west, the Adaville formation, gray sandstone and siltstone; 2) the Hilliard formation, dark-gray to tan claystone and siltstone; and 3) the Frontier formation, white to brown sandstone and dark-gray shale, with fossil oysters. Oyster Ridge, just east of Kemmerer, is composed of resistant, oyster-bearing sandstones of the Frontier formation.

Fossil Basin is mid-way between Kemmerer and Sage Junction. It is a long depression in the slab that moved east on the Absaroka overthrust fault. The Green River formation, known for its fossil fish, crops out through the basin, blanketing the older overthrust structures.

The Green River formation was deposited in a large, shallow lake that covered all of southwestern Wyoming approximately 55 to 50 million years ago. The lake slowly filled with delta deposits, chemical precipitates such as trona, and layers of oil shale. A variety of fish thrived in this warm lake, including catfish, herring, bass, gar and paddlefish. Fossil Butte is world famous for its millions of perfectly and delicately preserved fossil fish. Other fossilized organisms include lizards, crocodiles, turtles, snakes, insects, birds, and a variety

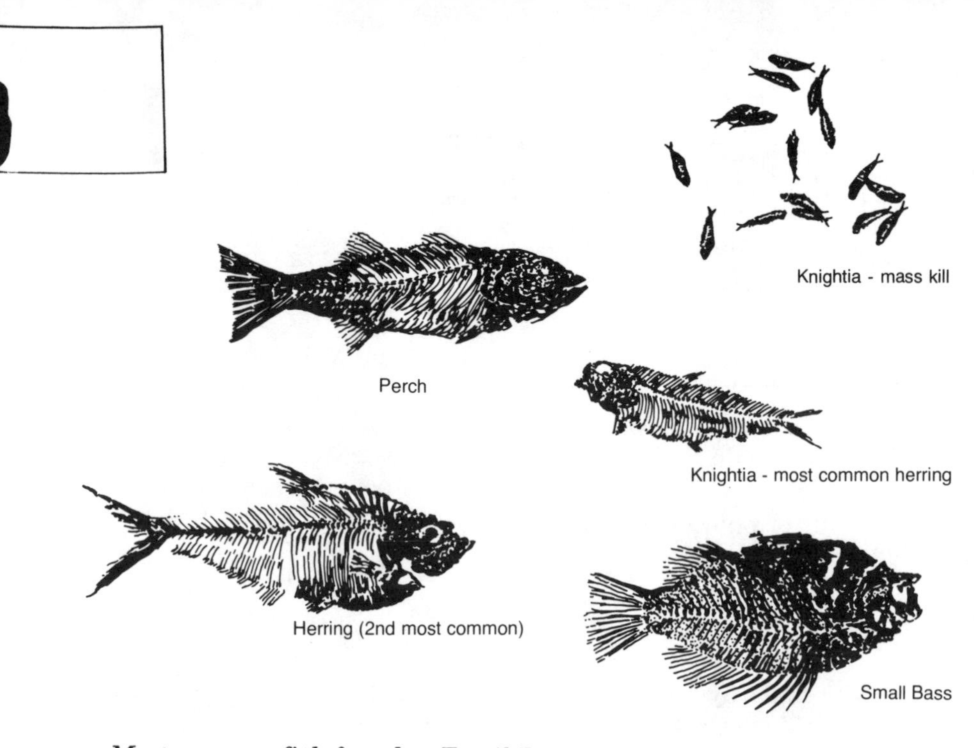

Most common fish found at Fossil Butte National Monument. –Adapted from McGrew and Casilliano

of plants such as palm fronds, fern, fig and cypress leaves. The fossils tell of a warm and subtropical climate, quite a contrast to today's harsh Wyoming climate! The visitor's center at Fossil Butte is well worth a stop. You can buy fossil fish specimens in Kemmerer.

In addition to fossil fish, the Green River formation is famous for its oil shale, a fine-grained sedimentary rock full of a kind of organic matter called kerogen that was chiefly derived from aquatic organisms and plant debris. Oil shale is not really a true shale; it is a dolomitic marlstone. There has been much debate about its origin.

Some geologists believe the oil shale formed as an organic ooze that did not decay because of a lack of oxygen at the bottom of the lake. That may also explain why the fossil fish are so delicately preserved. Other geologists believe the oil shale was deposited during times of extreme evaporation in a very shallow, playa lake. Regardless of its origin, the organic matter was converted into kerogen by time, pressure, and heat.

When heated to around 480 degrees in a retort, essentially a large pressure-cooker, the kerogen in oil shale vaporizes. The vapor condenses to form a thick oil that can be refined into useful hydrocarbon products. The Green River formation in parts of southwest Wyoming,

Eocene Green River formation at Fossil Butte National Monument.

northwest Colorado, and adjacent Utah contains large quantities of Eocene oil shale. Several problems have prevented full-scale mining of oil shale in these states: waste disposal, air quality, limited amounts of water for mining and processing, and competition with low-priced, imported oil.

The Tunp overthrust fault is a small overthrust sandwiched between the Absaroka overthrust to the east and the Crawford overthrust to the west. The trace of the Tunp overthrust is exposed aproximately 10 miles east of Sage Junction; the rocks in the slab that moved on the fault form the Tunp range and vary in age from Triassic, at the trace of the Tunp thrust, through Cretaceous near Sage Junction. Younger Tertiary strata, such as the tan mudstone, brown oil shale and siltstone of the Eocene Green River formation, locally overlie these older bedrock units.

US 89/30
Sage Junction—Cokeville 19 mi./30 km.

US 89/30 follows the Bear River Valley from Sage Junction to Cokeville and parallels the trace of the Crawford thrust fault. The fault is not well exposed along this section because the soft bedrock weathers readily and lies beneath a cover of young river deposits. The lower Cretaceous Sage Junction formation, gray and tan siltstone and sandstone, with minor limestone, conglomerate and coal, is exposed in the hills east of the road. Early settlers mined the coal for coke, hence the name Cokeville.

A spectacular anticline is exposed above the Crawford thrust fault immediately east of Cokeville, in a cut made by the Smith's Fork River. The folded limestone and sandstone belongs to the Pennsylvania Wells formation, and the dark gray chert and black shales belong to the Permian Phosphoria formation. The Crawford thrust fault is the fourth major thrust, counting from the east, in the Wyoming overthrust belt and is the westernmost major overthrust in the state; other major overthrust faults exist farther west, but they are not exposed in Wyoming. The Crawford thrust fault extends northward from the Crawford Mountains in southeastern Idaho.

Cokeville—Alpine Junction
87 mi./139 km.

Between Border, north of Cokeville, and Geneva, at the junction with US 89 to Montpelier, the road parallels the axis of the Sublette anticline. The east limb of the fold is in the hills east of the highway, where Pennsylvanian and Permian limestone and sandstone strata are exposed. The crest and west limb of the anticline have been faulted down to form the Thomas Fork Valley.

Between Geneva and Afton, the road winds and climbs over a divide formed by the thrusted and folded Jurassic and Cretaceous sedimentary rocks. The divide separates the Thomas Fork and Bear River drainages from the Salt River drainage. There are many individual formations along this section of road, rock types ranging from limestone to shale and sandstone. The greenish-gray limestone beds on the south side of the divide belong to the Jurassic Twin Creek formation. The bright red and maroon siltstone and shale beds intermittently exposed along the road are the Jurassic Preuss formation, which also contains salt and gypsum, and the lower Cretaceous Gannet group.

Between Afton and Alpine Junction, US 89 follows the Salt River through the Star Valley. The Salt River is not really salty; it gets its name from bedrock salt deposits and saline springs in the area. The Star Valley was named in 1870 by a Mormon settler, Moses Thatcher, who called it "...the star of all valleys." The Star Valley is well known for its dairy products, especially cheese.

Geologically, the Star Valley is a block dropped along the Grand Valley normal fault, which crops out at the base of the Salt River range on the east side of the valley. The Salt River range rose along the same fault.

The Grand Valley fault is seismically active. Fresh looking fault scarps at the base of the range testify to its on-going motion. Triangular faceted spurs, also evidence of active normal faulting, are well exposed east of Afton along the base of the range front. Faceted spurs are erosional remnants of a fault plane that has been dissected by streams to form a triangular outline; they only exist along mountain fronts experiencing active faulting.

Between Afton and Thayne, the highway goes through a narrow part of the valley that is formed by outcrops of Triassic and Jurassic strata west of the road, Pliocene sands and silts east of the road. These Pliocene sediments are collectively called the Teewinot formation. They were deposited by streams and in lakes within the Star Valley during its early stages of development.

US 26/89
Alpine Junction and Snake River Canyon—Hoback Junction
23 mi./37 km.

This highway segment cuts through two major thrust faults of the Wyoming overthrust belt and follows the beautiful canyon of the Snake River, sometimes called the Grand Canyon of the Snake River.

The west entrance to the Snake River Canyon is abruptly formed by the Grand Valley normal fault. This young, active fault dropped the Star Valley, also called Grand Valley, to form a northwest-trending graben along the Idaho-Wyoming border. After leaving its canyon, the Snake River turns sharply to the northwest and follows this valley into southeastern Idaho where it has been dammed to form Palisades Reservoir. The Grand Valley normal fault reactivated an old thrust fault in the subsurface similar to the Hoback fault (see Hoback Junction to Jackson section), in this case, the Absaroka thrust.

Massive cliffs along the western two-thirds of the Snake River Canyon, between Alpine and the Elbo, are gray limestones and white dolomites of Cambrian through Mississippian age. These Paleozoic marine sedimentary rocks were thrusted up along the Absaroka fault, the third major thrust fault, from the east, in the Wyoming portion of the overthrust belt. Rocks above the Absaroka thrust sheet form the Salt River range, which extends many miles to the south and north of here and is called the Snake River range in Idaho. The Absaroka thrust fault is very long and continuous, extending approximately 250 miles (400 km) from the north flank of the Uinta Mountains into southeastern Idaho. Most of the large oil and gas fields in the overthrust belt of southwestern Wyoming are in the Absaroka thrust sheet.

Two formations within the Absaroka thrust sheet stand out more than others in the Snake River Canyon; the Mississippian Lodgepole limestone forms cliffs of thinly bedded tan to gray limestone and the Ordovician Bighorn dolomite, which appears closer to Alpine, forms white to light gray cliffs in which you can see little bedding. The eastern one-third of the canyon cuts through Mesozoic sedimentary rocks below the Absaroka thrust that form low hills because they erode easily.

At the Elbo (look for Elbo Campground), the Snake River makes an abrupt turn. Upstream, between the Elbo and Astoria Hot Springs, the Snake flows south in a broad valley eroded in soft, red and maroon shales of Triassic age. Watch for the roadcuts along the west side of the highway. Downstream, between Elbo and Alpine, the Snake has cut a deep canyon through layers of much harder sedimentary rock. Why did the river leave its southerly trend through easily eroded Triassic shales? If you were a stream you'd probably follow the path of least resistance, rather than cut through harder bedrock formations.

The explanation for the stream's unlikely choice may be stream superposition, a concept developed by John Wesley Powell during his daring float trip down the Green and Colorado rivers in 1869, long before he became the second director of the U.S. Geological Survey. Superposition occurs when a stream cuts down through a cover of younger sedimentary rocks into folded and faulted terrane, like the overthrust belt. An alternate explanation proposes that the Snake River's abrupt course change to the west, where the overthrust belt bends to the northwest, might reflect the stream's tendency to erode its valley along the trend of fractures in the bedrock.

Astoria Hot Springs lies very near the Darby thrust fault, which is exposed on the sage hills north of the hot springs. Ground water migrates thousands of feet into the subsurface where it is warmed by the Earth's natural heat. Then the hot water moves upward along fractures until it eventually meets the Darby thrust and is discharged to the surface.

The Darby thrust fault is the second major overthrust in this part of Wyoming, between the Absaroka thrust in the the Snake River Canyon and the Jackson thrust adjacent to the town of Jackson. Thrust faults typically move older rocks over younger rocks. The Darby thrust placed older, well-bedded, cliff-forming sandstones of the Pennsylvanian Wells formation over younger Jurassic limestones, shales and sandstones. The block that rose along the Darby thrust, the Wyoming range, is composed of resistant Paleozoic limestone and sandstone strata. The Darby thrust extends from the north flank of the Uinta Mountains to southwest of Victor, Idaho, a distance of approximately 200 miles (322 km). It forms the frontal thrust fault of the overthrust belt in southwestern Wyoming, but near Big Piney it is offset to the west between the frontal Jackson-Prospect thrust and the Absaroka thrust.

US 189
Hoback Basin Rim—Hoback Junction

You will see two spectacular examples of Wyoming geologic structures between The Rim and Hoback Junction. The southeast-half of the route traverses the northern Green River Basin and offers spectacular views of the Gros Ventre and Wind River mountains. These mountains rose along thrust faults that cut very deep into the continental crust to bring ancient Precambrian basement rocks to the surface. The northwest half of this road segment cuts through the mountainous Hoback River Canyon, sliced through thrust faults and folds of the overthrust belt. The Wind River and Gros Ventre ranges are examples of foreland style or thick-skinned thrusting, while thrust faults in the Hoback River Canyon are thin-skinned or Sevier-style overthrust faults. These two styles of thrust faulting are uniquely juxtaposed in western Wyoming and this road segment offers a good view of both styles of faulting, as well as spectacular scenery.

The Rim is a high ridge of resistant conglomerate and sandstone of the Eocene Pass Peak formation, well exposed in roadcuts on its west side. The Rim forms a drainage divide: the Hoback Basin is the headwaters of the Hoback River which flows into the Snake and

Flying Buttress Mountain along the southwest flank on the Gros Ventre Range from the Granite Creek Road, which intersects highway 189 in the Hoback River Canyon.

Columbia rivers, which empties into the Gulf of California. Structurally, the Hoback Basin and The Rim are all part of the northern Green River Basin, a large sediment-filled basin between the overthrust belt and the Gros Ventre and Wind River mountains.

Between The Rim and Battle Mountain, the road follows Fisherman Creek and the upper Hoback River. As you drive through the Hoback Basin, you see outcrops in the low hills on either side of the road of the Eocene Wasatch formation, vari-colored shale and sandstone, and the Eocene Pass Peak formation, yellow-tan sandstone. Both are stream and floodplain sediments that were deposited approximately 55 million years ago during final stages of thrusting in the overthrust belt to the west, and final stages of uplift of the Gros Ventre Range to the northeast.

The Gros Ventre Range frames the northeast margin of the Hoback Basin, foreland style uplift of Precambrian basement rocks and Paleozoic-Mesozoic strata. The range is bound on its southern margin by the Cache Creek thrust fault, which cut deep into the crust to bring the ancient gneisses and schists in the core of the range to the surface. This thrust fault was active in the late Paleocene and early Eocene time, approximately 55 to 60 million years ago, during the Laramide mountain building event.

Battle Mountain, immediately east of the Granite Creek Road, marks the trace of the Jackson-Prospect thrust fault and is the lead-

Battle Mountain, formed by Jurassic Nugget sandstone thrust over the Paleocene Hoback formation, marks the trace of the Prospect thrust fault in Hoback Canyon along Highway 189.

ing edge, or eastern margin, of the Wyoming overthrust belt. Here, the Jurassic Nugget sandstone was thrust eastward over the drab, grayish tan shales and sandstones of the Paleocene Hoback formation. The US Forest Service campground on the east side of Battle Mountain offers an excellent view of the Jackson-Prospect thrust fault and Battle Mountain (Kozy Campground). This is a significant boundary in western Wyoming, marking change in structural style; overlapping thrust faults of the overthrust belt lie to the west; to the east lie foreland style uplifts of basement rocks, such as the Gros Ventre and Wind River mountains.

The Hoback River Canyon slices through a complicated array of overthrust faults and folds that raised Paleozoic and Mesozoic sedimentary rocks to form the Hoback Range—the Hoback River cut through its northern end. At the east end of the canyon, red, maroon, pink, and yellow sandstones and shales of Mesozoic age are exposed. In particular, look for the salmon-colored sandstone beds near the Granite Creek Road intersection at Battle Mountain; this unit is the Jurassic Nugget sandstone, an important reservoir rock for oil and gas in the overthrust belt of southwestern Wyoming and northeast Utah. At the west end of the canyon, thousands of feet of well-bedded, gray to tan limestone and dolomite strata, mostly of Mississippian and Denovian age, are visible.

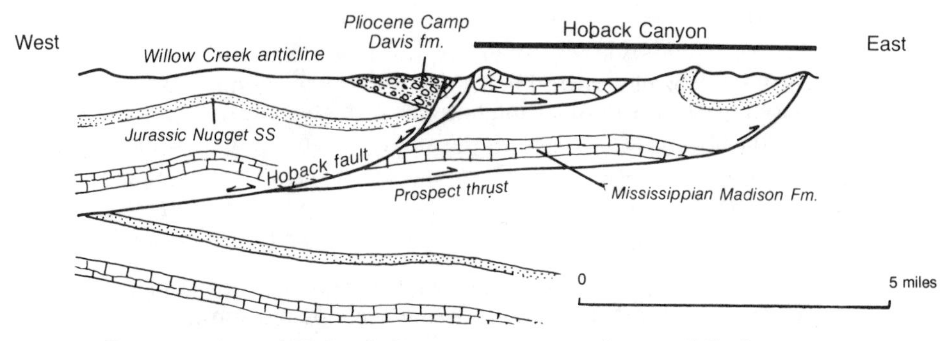

Cross section of Hoback Canyon area, southeast of Jackson, Wyoming. –Adapted from Royse, Warner and Reese (1975, p.46)

The west end of the Hoback River Canyon is marked by the trace of the Hoback normal fault. A warm sulfurous spring flows from the fault into the Hoback River. An excellent pull-off area allows motorists to inspect the warm spring and Hoback River.

The road follows low sage-covered hills between the west end of Hoback Canyon and Hoback Junction. The hills are composed of Cretaceous sandstones and shales overlain by the Pliocene Camp Davis formation, named for Camp Davis, the University of Michigan Geological Field Station located in this valley. The younger Camp Davis formation is a conglomerate that was deposited in this valley as it dropped along the Hoback normal fault. Good outcrops of the Camp Davis formation are on the hillside behind the Spotted Horse Ranch.

US 191
Hoback Junction—Jackson
13 mi./21 km.

Between Hoback Junction and Jackson, US Highway 191 follows Flat Creek and the Snake River through South Park, a southern extension of Jackson Hole. Like Jackson Hole proper, South Park has been dropped along faults to form the valley. The main controlling fault is the Hoback normal fault, on the east side of the valley—the highway follows the trace of this fault! The Hoback normal fault reactivated an older thrust fault within the Jackson-Prospect thrust sheet, as described in the chapter's introduction.

Approximately half-way between Hoback Junction and Jackson, the road crosses the Snake River. If you look to the east near here, you can see the old highway on a steep hillside. The bedrock on this hill is poorly consolidated Cretaceous shale of the Aspen formation and is very unstable and subject to slumps and slides. The highway had to be relocated to its present position because repeated slides destroyed the old road. This is a good example of how geology can affect very practical, human activities. These Cretaceous shales are exposed in roadcuts near Hoback Junction—look for the black and gray-green shale outcrops adjacent to the road and along the banks of the Snake River.

The geological setting of Jackson deserves special emphasis. This is an area of complicated, spectacular, dynamic geology. The town of Jackson is in a valley formed by two thrust faults. In a sense, the valley has been caught between the jaws of a giant vise that trends east-west. The vise's northern jaw is the Cache Creek thrust fault; that to the south is the Jackson-Prospect thrust fault. The valley is between these two thrust faults, a unique structural setting, because the two thrust faults moved in opposite directions towards one another at the same time, and they represent two different styles of thrust faulting. The Cache Creek thrust moved southwest in early Eocene time, uplifting Precambrian basement rocks in the core of the Gros Ventre range. At roughly the same time, the Jackson-Prospect thrust moved northeast carrying Paleozoic sedimentary rocks in its hanging wall, but no Precambrian basement rocks, as this was a thin-skinned detachement thrust fault. The rocks on east and west Gros Ventre Buttes on the north side of town are remnants of the Cache Creek thrust, while the sedimentary rocks of Snow King Mountain south of Jackson are remnants of the Jackson-Prospect thrust sheet. As you can imagine, the valley that Jackson sits in was nearly squeezed between these two large thrusts!

View of the Teton Range from Rendezvous Mountain. Grand Teton is the highest peak on skyline (13,770 feet).

Profile of Tetons and Jackson Hole. –Adapted from Love and Reed (1971)

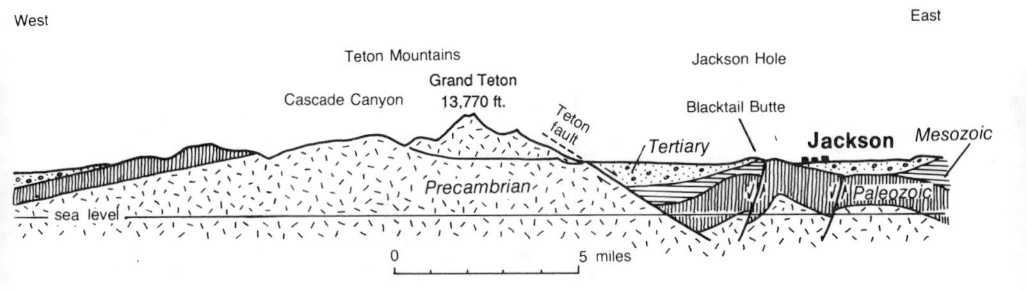

VII
Grand Teton National Park

INTRODUCTION

In many respects, the Teton Range is the jewel of Wyoming. Although the Tetons are a relatively small mountain range by Rocky Mountain standards, about 40 miles long by 15 miles wide, their sheer rise above the valley floor and rugged, glacier-carved topography make them one of the most spectacular ranges in North America. The Tetons form a giant wall of stone, separating the high mountains and basins of Wyoming from the volcanic lowland of the Snake River Plain to the west.

The range was first named "les Trios Tetons," or "the three teats," by French fur trappers and explorers who observed the Grand, Middle, and South Teton peaks from the west. Indians called the range "tee-win-ot," meaning three pinnacles, and one peak retains this name. In the 1870s a group of geologists, associated with the U.S. Geological and Geographical Survey of the Territories, attempted to rename the Grand Teton "Mount Hayden" after their director, Dr. F.V. Hayden, but fortunately the name Grand Teton and Teton Range stuck.

The town of Jackson, Jackson Hole and the Grand Teton Range from Snow King Mountain.

Like other geological features in Wyoming and elsewhere, the Teton Range is a result of the constant struggle between internal Earth processes acting to uplift the crust, and surficial processes of weathering and erosion acting to smooth-out and level-off the Earth's surface. In some parts of the U.S., weathering and erosion dominate and the Earth's surface is smooth and flat; in other places, like the Rockies, uplift and mountain building dominate. The Teton Range is unique because the scale of both uplifting and erosional processes is so spectacular! Why does the Teton Range exist in the first place, and why does it look the way it does—both uplift and erosion are important.

Uplift of the Teton Range

The Teton Range is the youngest on the Wyoming landscape. These mountains are less than 10 million years old, and still rising! Most of Wyoming's mountains, like the Wind River, Beartooth, Bighorn, Laramie, and Medicine Bow ranges, are around 60 million years old, formed by compressional stresses and contraction of the crust during the Laramide orogeny. Most of these older ranges are bounded by reverse faults and

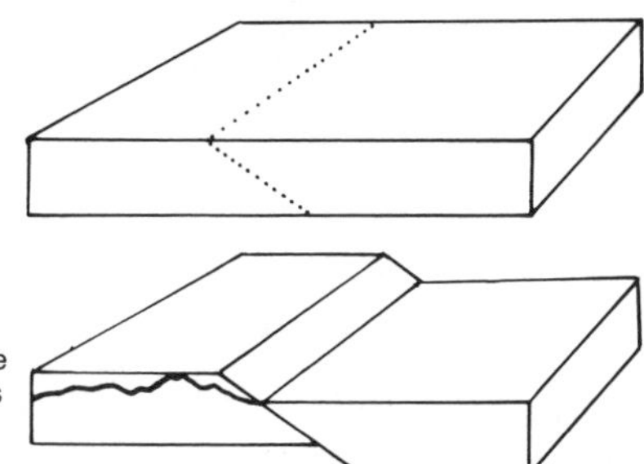

Block diagram of a normal fault, showing Teton fault as an example.

thrust faults that result from shortening of the earth's crust. The Teton Range is the product of crustal extension—a geologic process occurring today that is pulling the crust apart in an east-west direction.

Crustal extension has produced the Basin and Range Province of mountains and valleys from western Wyoming and southern Idaho, through Utah and Nevada and as far as southern Arizona and southeastern California. Mountain ranges in the Basin and Range are bounded by normal faults that pull the crust apart when they move, while at the same time vertically offsetting the crust into mountains and valleys to create block-faulted mountains.

The east flank of the Teton Range is bounded by an enormous normal fault that raised the Teton Range and dropped Jackson Hole. This fault dips to the east and may have as much as 25,000 feet of displacement—almost 5 miles! The Teton fault crops out at the "break in slope," where the steep mountain slope meets the valley floor, at the eastern base of the range. The fault can be identified by "fault scarps," or steep slopes, at the base of Mount Teewinot and Mount St. John. Movement on the Teton normal fault continues to raise the Tetons, while dropping Jackson Hole. As you would expect, the Teton fault is seismically active, but most earthquakes are too small for people to notice. However, the potential does exist for a sudden displacement on the Teton fault accompanied by a large earthquake. It is impossible to estimate when such a large earthquake might strike.

The Teton fault is the easternmost normal fault separating the actively extending Basin and Range to the west, from the unextended Rockies to the east. If the Basin and Range Province continues to grow by crustal extension and thinning, it is possible that more and more of western Wyoming could be "eaten-up" by normal faults as new block-faulted mountain ranges and valleys develop. This, however, is pure speculation and would take millions of years to happen.

We now have a mountain range and can turn to the geological processes of weathering and erosion that are carving magnificent canyons and peaks as they wear the range down.

Surficial Geological Processes

Weathering and erosion are primarily driven by water-related processes. The peaks are subjected to stream erosion, rockslides, and wind, but alpine glaciation is, and was, the dominant process responsible for shaping the myriad of spires, canyons, ridges, and basins within the Tetons. The Teton Range is unique and spectacular, not because it has been glaciated, for almost all mountains in the Rockies have been glaciated, but because of the tremendous magnitude of glacial carving!

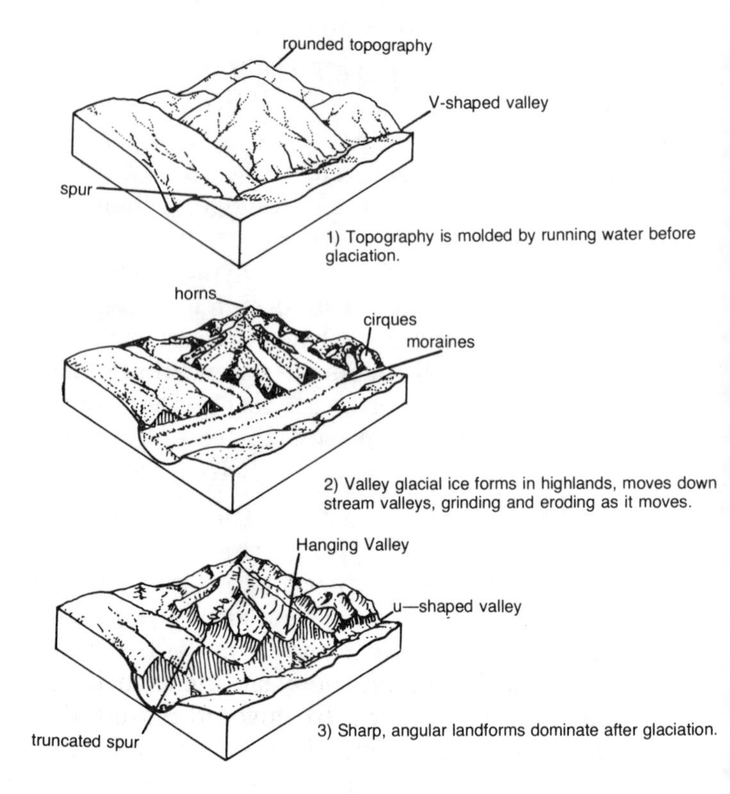

Some of the most spectacular scenery in the world is produced by valley glaciation. These diagrams show a landscape, formed by stream erosion, as it appears before, during and after glaciation. –Adapted from Hamblin (1985)

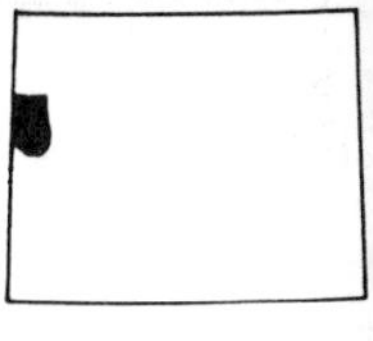

Glaciers form whenever more snow accumulates in the winter than melts in the summer. When the snow becomes deep enough, it recrystallizes into glacial ice. Under the pull of gravity, the ice on the side of a mountain begins to move downhill, gouging out the valley. Where glacially gouged valleys lie side-by-side, a sharp ridge or arete separates them. When glaciers carve canyons on all sides of a mountain, they leave a craggy horn peak, like Switzerland's Matterhorn and the Grand Teton. The head of a glacier lies in a bowl-shaped cirque, which may hold a tarn lake after the ice melts. Lake Solitude is a tarn. Hanging valleys form where a high glacier in a cirque basin flows into a larger valley glacier that has carved a deeper canyon. Where the glaciers flowed out of their canyons onto the floor of Jackson Hole, they shoved up huge piles of eroded debris, to form terminal moraines, like a pile of dirt in front of a bulldozer. These are natural dams for streams flowing out of the mountains, and create lakes like Leigh, Jenny, Bradley, Taggart, and Phelps. You can easily see these glacial features from almost any viewpoint in the park, but the road from Moose to Jenny Lake is especially good for "glacier gazing."

When did all this glaciation take place? Obviously it must have been in the recent geological past because the glacial features are so well preserved. The "Ice Age" or Pleistocene is the period of Earth history from about 2 million years ago to the present. It was a time of cooler climates, more precipitation, and widespread glaciation. Continental glaciers covered the land at high latitudes in the northern and southern hemispheres, and alpine glaciers covered mountain ranges everywhere. The causes of the Ice Age are not fully understood.

In the Rocky Mountains of Wyoming, at least three major glacial advances occurred in the last 250,000 years. The oldest and most widespread was the Buffalo glaciation of around 200,000 years ago. Thick tongues of ice converged on Jackson Hole from the Beartooth-Absaroka-Yellowstone region to the north, the Wind River and Gros Ventre ranges to the east, and the Teton Range on the west. Jackson Hole filled with more than 2,000 feet of ice. Ice covered Signal Mountain, Blacktail Butte, the Gros Ventre Buttes, and Snow King Mountain. Ice flowed south along the east face of the Teton Range, then

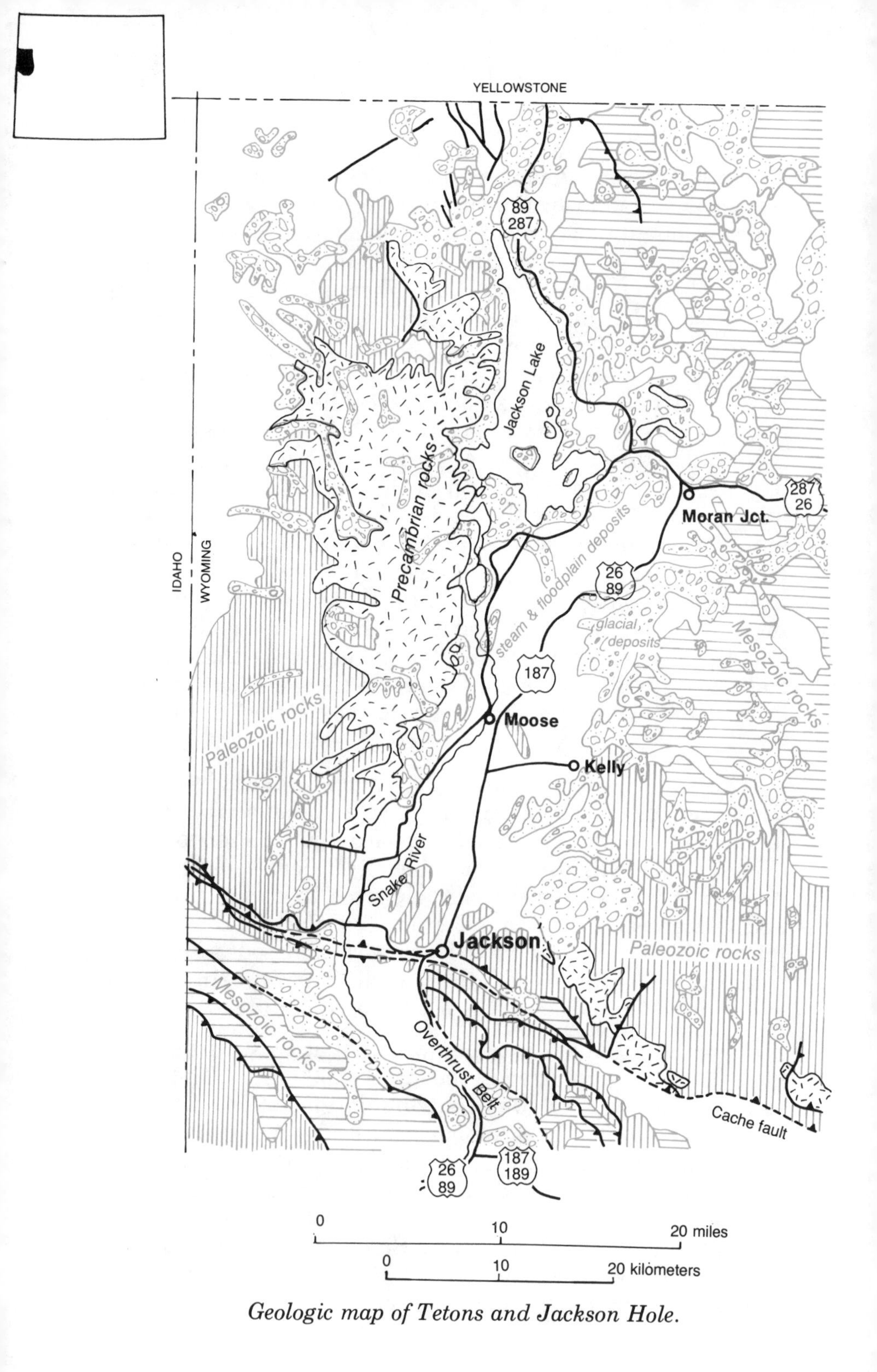

Geologic map of Tetons and Jackson Hole.

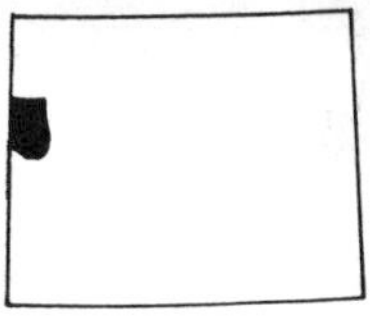

through the Snake River Canyon to Idaho. Let your imagination go and picture that enormous river of ice slowly creeping down the valley.

The second glaciation was the Bull Lake glaciation, between 130,000 to 200,000 years ago. Again, a large tongue of ice moved south down the Buffalo River Valley from the Absaroka-Yellowstone region and joined ice from the Tetons on the floor of Jackson Hole. Remnants of Bull Lake till, or morainal debris, include Timbered Island south of Jenny Lake and Signal Mountain. The type locality of the Bull Lake event is at Bull Lake on the northeast side of the Wind River Mountains where excellent moraines are preserved.

The most recent glacial event is named after Pinedale, Wyoming, where these young moraines dam up Fremont Lake. The Pinedale event probably began around 70,000 years ago and ended 15,000 to 20,000 years ago in the Rockies. This last event is most noticeable in Jackson Hole and the Tetons, for it left in its wake the lakes, rugged peaks, even the sagebrush-covered, gravel flats of the valley. Pinedale glaciers flowed down the canyons to the Teton Range and spilled onto the floor of Jackson Hole, creating large terminal moraines that became natural dams for the many lakes along the eastern base of the range.

Valley glaciers from the northern Tetons, and perhaps the Yellowstone Plateau, coalesced on the floor of northern Jackson Hole to form a large, continuous sheet of ice called a piedmont glacier at the site of Jackson Lake. The southern extent of this ice is marked by timbered moraines that span Jackson Hole, collectively known as the Burned Ridge moraine. Outwash gravels, spread by sediment-choked streams, extend south of the Burned Ridge moraine complex, creating the extremely flat surface of Jackson Hole. As the Jackson Lake glacier melted back from Burned Ridge to the present site of Jackson Lake, chunks of ice were left stranded in the outwash gravels, where they left small depressions, kettles, as they melted. They are locally known as "The Potholes." The ice remained stationary over the present site of Jackson Lake for many years and built up a substantial moraine, creat-

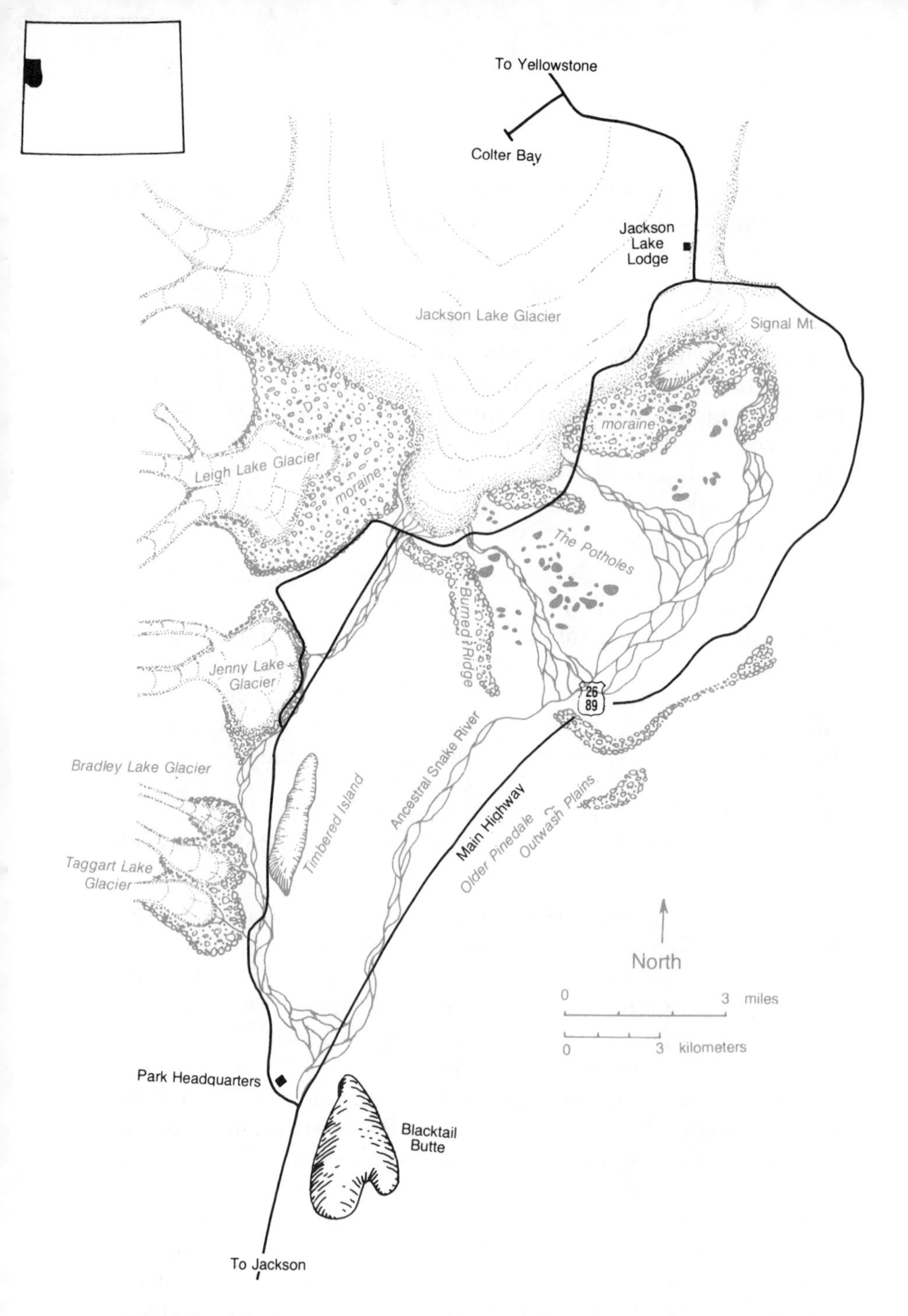

Glacial geology of Jackson Hole, at the time Pinedale glaciers deposited their moraines. Jackson Lake was dammed by moraines at the south end of Jackson Lake glacier. –Adapted from Love and Reed (1971)

ing a natural dam for Jackson Lake. The Jackson Lake we see today has been enlarged by the man-made dam at the lake's east end.

Although most of the glaciers that carved out the Tetons melted 15,000 to 20,000 years ago, a few hang on in deep, shaded canyons. Examples you can easily see include Skillet Glacier on the east face of Mount Moran, Falling Ice Glacier on the southeast face of Mount Moran, and the Teton Glacier at the base of the east face of the Grand Teton.

The magnificent mountains of the Teton Range and the floor of Jackson Hole bear a strong imprint of their glacial heritage. Elk and deer enjoy the wide, flat expanses of sage-covered flats punctuated by timbered glacial moraines on the floor of Jackson Hole, while moose and bear enjoy the cooler recesses of u-shaped alpine valleys and cirque basins. It is all the product of geology and time!

US 191/89/26
Jackson—Gros Ventre Junction
7 mi./11 km.

US 191/89/26 from Jackson north to the park entrance follows the base of East Gros Ventre Butte. This prominent hill is composed of Tertiary andesitic volcanic rocks overlying Upper Paleozoic limestone and sandstone beds of the Madison through Phosphoria formations. Buffalo glacial ice that came from Yellowstone Park approximately 200,000 years ago completely buried the butte.

The floodplain of Flat Creek and the National Elk Refuge is on the east side of the road, with Gros Ventre Mountains in the distance. The Gros Ventre Range is more than 50 million years older than the Teton Range. It rose during the Laramide orogeny by compression and shortening of the crust.

During the winter months, 5,000 to 10,000 elk make their home on the National Elk Refuge and are fed with over 5,000 tons of hay. The Indian name for elk is "wapiti."

As the road climbs above the Flat Creek floodplain, the Teton

Range boldly rises into view. The southern Tetons are capped by Paleozoic sedimentary rocks. For example, the summit of Rendezvous Mountain, at the top of the Jackson Hole Ski Tram, is Devonian carbonate rocks of the Darby formation. The sedimentary rocks on the skyline of the southern Teton Range have inspired the name "gabled peaks." Due east is Sheep Mountain or the "Sleeping Indian" the south peak is the Indian's face; the north peak is his rounded chest. Sheep Mountain is an uplift of Mississippian Madison limestone on the northeast side of the Gros Ventre Range.

As you approach Gros Ventre Junction, the road crosses a flat, sage-covered alluvial terrace of the Snake River, which flows through the ribbon of green beyond the sage flats. Blacktail Butte is the prominent hill in the middle of Jackson Hole, just east of Moose Junction. It is composed of Paleozoic through Tertiary sedimentary rocks and, like the Gros Ventre Buttes, was completely covered by Buffalo glacial ice 200,000 years ago.

Gros Ventre Junction—Gros Ventre Slide 11 mi./18 km.

The road to the Gros Ventre (pronounced "grow vont") Slide follows the old route from Jackson Hole over Union Pass; this pass was used by Indians and mountain men before the present routes into Jackson Hole were established. The Gros Ventre River drains the north flank of the Gros Ventre Range. The name Gros Ventre means "big belly," a name given to the local Indians by early French fur trappers. Old Indian camps are scattered along the river.

The Gros Ventre landslide was triggered by an earthquake on June 23, 1925, after several weeks of heavy rain coupled with a late snow melt, which made the ground soggy and unstable. The slide was observed first-hand by Guil Huff, a local rancher, who was looking for stray cattle. It happened at 4:20 PM, and Mr. Huff narrowly escaped at a full gallop.

In 3 minutes, some 50 million cubic yards of rock surged down the side of Sheep Mountain, then 400 feet up the opposite bank, creating a dam some 2000 feet wide, a mile long, and 225 to 250 feet high across the Gros Ventre River. State Surveyor W.O. Owen estimated that if

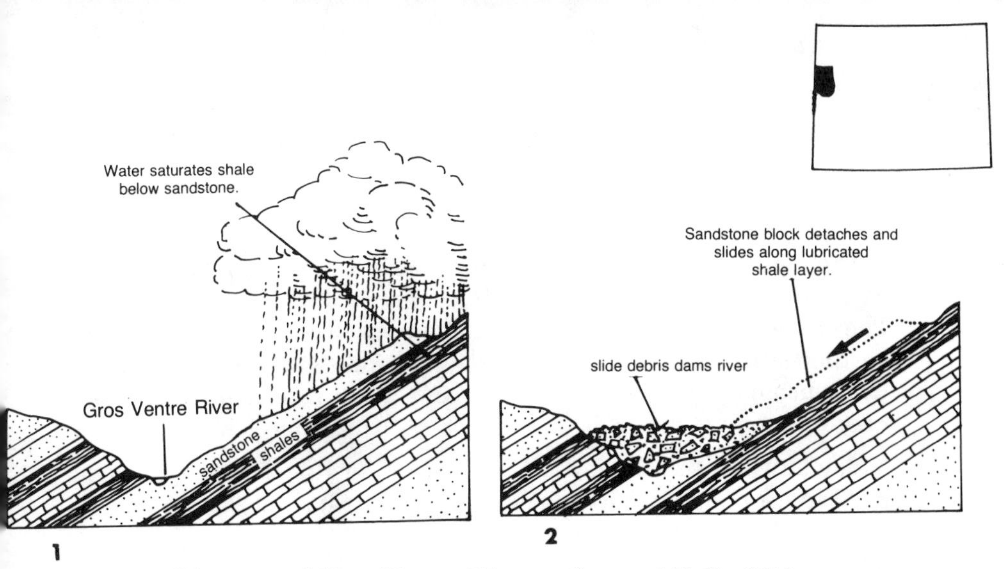

Diagram of Gros Ventre River valley and Kelly Slide.

the builders of the Panama Canal could have moved earth with the speed of the Gros Ventre Slide, the entire canal could have been completed in 54 minutes!

On May 18, 1927, nearly 2 years after the slide, the dam across the Gros Ventre River failed as high water went over the top. The resulting "Kelly flood" killed 6 people and hundreds of livestock, left 40 families homeless, and covered the fertile Gros Ventre River floodplain with gravel and boulders. Slide Lake is a remnant of the former lake.

The Gros Ventre landslide, or more correctly "rockslide," was not unique in this area. If you drive farther up the Gros Ventre River you can see previous landslides that are now tree covered.

Four conditions combined to bring the earth down on that June day in 1925: 1) the layers of sedimentary rock in this area dip or tilt 15 to 20 degrees toward the Gros Ventre River; 2) layers of sandstone, called the Tensleep formation, rest upon a slippery shale layer; 3) the sedimentary rock layers were undercut by the Gros Ventre River, as it meandered across its valley, removing support at the base of the tilted beds; 4) the late snow melt and heavy rains saturated the soil and infiltrated the bedrock, reducing friction on the bedding planes and making the underlying shale layer more slippery. During the earthquake tremor, the Tensleep sandstone could no longer maintain its inclined attitude on the slippery shale because it had no support at the bottom of the slope, and the mountain came tumbling down. The situation was inevitable; it happened in the past, and will happen again when conditions are right.

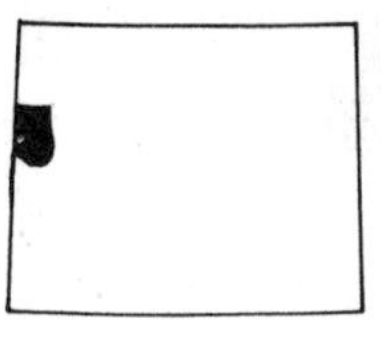

US 191/89/26
Gros Ventre Junction—Moose
9 mi./14 km.

The sage-covered flats on this section of road are composed of Snake River alluvium, or river sediments, and reworked glacial outwash. The gravelly soil does not retain water very well, so trees do not grow on the flats. Sagebrush is ideally suited for these soil conditions. Uinta ground squirrels, sage grouse and coyotes are common on the sage flats.

Blacktail Butte is east of the road at Moose Junction. Its position right in the middle of Jackson Hole is puzzling; it is probably a small, uplifted, fault-bound block.

Moose—Jenny Lake
8 mi./ 13 km.

This is perhaps the most scenic road in Grand Teton National Park. From the park entrance to Moose, the road immediately climbs up a gravel terrace of the Snake River. The Tetons soar over 7000 feet above the valley floor! Bradley and Taggart lakes, at the base of the range, are natural lakes formed by glacial moraine dams; the morainal debris contains gravel, boulders, and clay, which retains groundwater and enables lodgepole pine to grow. Timbered Island forms a ridge east of the road between Bradley Lake and Jenny Lake. This hill is an erosional remnant composed of morainal debris from pre-Pinedale glaciations.

Glacier Gulch viewpoint, right across from Timbered Island, is a fantastic place to study glacial features. The Teton Glacier lies in a cirque at the head of Glacier Gulch, shaded by ridges of the Grand Teton, Mt. Owen and Mt. Teewinot. It is about 3,500 feet long and 1,200 feet wide. Ice in the central part of the glacier moves about 30 feet per year.

Mount Teewinot is the prominent peak that comes right down to the south end of Jenny Lake. Look along the base of this mountain for "fault scarps," which indicate past displacement on the Teton normal fault. These east-facing scarps are best seen in the late afternoon light when a shadow is cast across the face of the scarp. Fault scarps are relatively common in the Basin and Range Province.

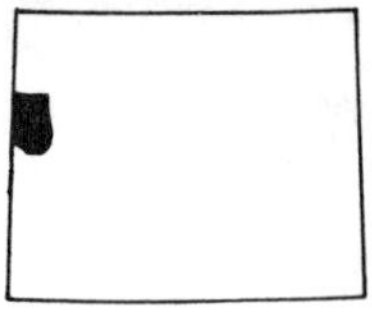

South Jenny Lake— North Jenny Lake Junction Jenny Lake Loop Drive

5 mi./8 km.

From South Jenny Lake Junction, the road heads straight northeast over sagebrush flats composed of outwash and reworked alluvial gravels. Notice how the terminal moraine around Jenny Lake is timbered with lodgepole pine, again because the high clay content of the soil retains more water than the gravelly, alluvial outwash flats.

At North Jenny Lake Junction turn southwest towards Jenny Lake. The first part of the road traverses morainal debris of Pinedale age that surrounds Jackson Lake and was derived from the Jackson Lake piedmont glacier.

Stop at the Cathedral Group viewpoint for a spectacular view—Mount Teewinot is on the left (12,325 feet), Mount Owen on the right (12,928 feet), and the Grand Teton in the middle (13,770 feet). These three peaks form the highest part of the Teton Range, topographically and structurally, and have the greatest displacement on the Teton normal fault. At the base of Mount St. John, due west, you will see another large fault scarp formed by displacement on the Teton normal fault. Also, to the northwest you can see Falling Ice Glacier clinging to the side of Mount Moran; notice the black dike of igneous rock that extends from the top of the glacier to the summit. This dike and others in the Tetons are composed of diabase, a dense igneous rock similar in composition to basalt. These dikes intruded cracks in the Precambrian basement about 1.2 billion years ago, long before shelled organisms evolved in the ocean.

As you drive south on the Jenny Lake terminal moraine, look for the large boulders that are typical of glacial debris; these were plucked from the high canyons and carried to the base of the range by glacial ice. They are called erratics.

Jenny Lake is one of the most picturesque spots in the Rocky Mountains. It was named after the Indian wife of "Beaver Dick Leigh," a local trapper and guide. Jenny Lake is the second largest lake in the Park. The elevation of the lake surface is 6,783 feet. It is 256 feet deep and formed as Cascade Creek flowed into the natural dam created by the Cascade Canyon terminal moraine. The view up

Cascade Canyon is spectacular and shows the gouged valley profile characteristic of glaciated valleys. The trail up Cascade Canyon offers the opportunity to study glacial landforms first-hand, and can be reached by hiking around the lake or taking the boat across. Many avalanche chutes are visible on the mountainside above Jenny Lake.

US 287/89
North Jenny Lake Junction—Signal Mountain—Jackson Lake Junction
9 mi./15 km.

Immediately past North Jenny Lake Junction, the road crosses the Burned Ridge moraine. This moraine wrapped around the south end of the Pinedale-age Jackson Lake glacier. You will also drive by "The Potholes," small depressions left by large chunks of ice that melted in the outwash debris of the Jackson Lake glacier. The correct geological term for this feature is "kettle," not pothole; a pothole forms by erosional scour on the bed of a stream.

The Mount Moran scenic turnout is in an old stream outlet channel through the Jackson Lake recessional moraine. From here there is an excellent view of Falling Ice and Skillet glaciers on the southeast and east faces of Mount Moran, respectively. Three other glaciers cling to the north flank of Mount Moran, but they are not visible from this point. The "black dike" is clearly visible. Mount Moran is capped by a small patch of Cambrian Flathead sandstone, left as an erosional remnant; look for the light-colored patch of rock on the summit. This patch of sandstone can be correlated by seismic reflection methods with the same formation beneath Jackson Hole; thus we estimate 25,000 feet of offset on the Teton normal fault.

The road traverses the Jackson Lake recessional moraine, of Pinedale age, as you approach Signal Mountain Lodge. Signal Mountain is composed of Bull Lake glacial moraine left as a remnant of this older glaciation.

A narrow, winding road leads to the top of Signal Mountain, and the view from the top is well worth the drive! Signal Mountain was completely burned in the 1880s; the lodgepole pine forest on the mountain has regrown naturally since that time and may eventually evolve into a mature spruce-fir forest.

Signal Mountain was completely overridden by ice during the Buffalo glaciation, around 200,000 years ago—gravel and cobbles left by this tremendous ice sheet can be found on the summit today. Evidence of more recent glaciations is visible from the south summit. To the south, the low, timbered Burned Ridge moraine spans the floor of Jackson Hole. This moraine marks the south extent of glacial ice that covered the valley less than 15,000 years ago—Pinedale glaciation. Kettles, depressions formed by chunks of melting ice, dot the valley floor. Blacktail Butte and East and West Gros Ventre buttes are visible to the south and are fault-bound uplifts. Like Signal Mountain, they were completely covered by glacial ice in the past. The Teton Range rises majestically to the west.

From Signal Mountain to US 287/89, the road follows the eastern shore of Jackson Lake on the Jackson Lake recessional moraine. Jackson Lake Dam was built in 1916 to provide downstream irrigation water and control spring flooding. The Bureau of Reclamation has extensively modified the dam to make it stronger in the event of an earthquake—remember, earthquake potential does exist here, caused by the active uplift of the Tetons!

After crossing Jackson Lake Dam, the road traverses Willow Flats which is floored with Jackson Lake ground moraine and reworked outwash gravels. This low, marshy area contains numerous beaver dams and is excellent moose habitat.

US 191/89/26
Moose—Moran Junction
19 mi./30 km.

Moose Junction started as a ferry crossing. The Menor Ferry, a flat-bottomed, homemade ferry, was operated by William Menor (pronounced "Meaner") at this spot for more than 25 years until the Snake River bridge was built in 1927. It cost 50 cents to cross the Snake River on the ferry, and if you happen to see the Snake at high water in the spring, you'll realize the price was a bargain! Blacktail Butte is the prominent hill directly east of the highway.

River terraces in glacial outwash gravels cut by the Snake River in Jackson Hole.

North of Moose Junction to Deadman's Bar the road is built on alluvial terrace gravels of the Snake River. Antelope Flats is the broad, sage covered, terrace surface extending northeast of Blacktail Butte. These gravels are reworked glacial outwash sediments that were deposited south of the Burned Ridge moraine. The Snake River cut its present channel into these gravels in the last 15,000 years, following the end of the Pinedale glaciation. The Snake River is carrying far less sediment than the ancestral Snake River carried in glacial times, when glaciers supplied much sediment, so it is capable of eroding a deeper channel. This is why many rivers in the central Rockies have cut terraces into former river alluvium.

The Gros Ventre Slide, on the north flank of Sleeping Indian Mountain, is visible across Antelope Flats to the east.

At Snake River Overlook, Deadman's Bar, you have an excellent view of the meandering Snake River. Deadman's Bar was the site of a triple murder in 1886, when John Tonnar killed his three companions while panning for gold.

From the Snake River Overlook the road drops down through the pine-covered Burned Ridge moraine of Pinedale age, to the Spread Creek drainage. Notice the glacial kettles within the moraine. Some of these have filled with water to form small, swampy lakes.

The low hill east of the road immediately north of Spread Creek is capped with remnant Bull Lake glacial till, resting on Cretaceous marine sedimentary rocks. From Spread Creek to the Buffalo Fork River the road crosses Quaternary alluvial sediments. The quartzite cobbles in the Buffalo Fork River were derived from the Late Cretaceous Harebell conglomerate which crops out upstream. The source of these quartzite cobbles in Cretaceous time is still a topic of controversy among geologists.

US 287/191/89
Moran Junction—Jackson Lake Junction
4 mi./6.5 km.

Moran Junction is the northeast entrance to Grand Teton National Park. It and Mount Moran are named after the famous landscape artist, Thomas Moran, who accompanied the US Geological and Geographical Survey of the Territories in the 1870s; Moran's paintings of Yellowstone, along with the photographs of W.H. Jackson, were instrumental in establishing that area as our Nation's first national park.

One mile south of Pacific Creek, the upper Cretaceous Harebell conglomerate crops out near the road on the side of Lozier Hill. This formation consists of quartzite cobbles interbedded with greenish sandstones and claystones. This formation was deposited above sea level and the conglomerate indicates there must have been a highland somewhere to the west, although the exact source has yet to be determined.

The oxbow bend of the Snake River is a classic example of a cut-off meander. Rivers have a natural tendency to meander back and forth in their floodplains. The greatest erosion invariably occurs on the outside of meander loops, making them larger. When two loops join, the meander is cut-off and an oxbow lake is left marking the former course of the river.

US 89/191/287
Jackson Lake Junction—
South Entrance of Yellowstone
21 mi./34 km.

Jackson Lake Lodge stands on a high ridge of Pinedale-age glacial moraine. This moraine is composed of gravel, sand, silt, and clay deposited along the front of the Jackson Lake piedmont glacier.

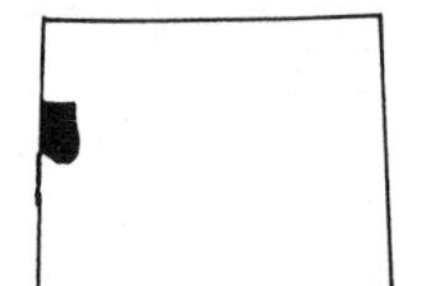

Glacial ice had melted out of the valley by about 15,000 years ago, leaving the moraine as a natural dam that impounded Jackson Lake; the original Jackson Lake was enlarged by the man-made dam at the Snake River outlet.

Emma Matilda and Two Ocean lakes are in the hills east of Jackson Lake Lodge. Two Ocean Lake is misnamed because it drains only to the Pacific Ocean via Pacific Creek, the Snake River and the Columbia. Two Ocean Creek, however, does flow along the Continental Divide and has two forks that separate; one drains to the Gulf of Mexico and the other to the Pacific Ocean.

The timbered hills around Colter Bay are composed of Pinedale glacial moraine. The road follows this morainal debris intermittently along the east shore of Jackson Lake.

Between the north end of Jackson Lake and the south entrance to Yellowstone National Park, the road crosses the Huckleberry Ridge tuff, glacial sediments, alluvial river sediments, and Cretaceous bedrock formations. The Huckleberry Ridge tuff is a tan, ash-flow deposit that was erupted 2 million years ago from a giant caldera in Yellowstone; it is composed of rhyolite. To the west, the north end of the Teton Range gradually looses relief and finally disappears beneath the young volcanic rocks of Yellowstone.

VIII
Yellowstone National Park

INTRODUCTION

Yellowstone National Park was established by a Congressional Bill in 1872 as this country's first national park. Dr. Ferdinand Vandiveer Hayden, geologist in charge of the U.S. Geological and Geographical Survey of the Territories, played a key role in establishing Yellowstone as a national park through his geological exploration of the region in 1871-72 and his subsequent publications. To avoid commercial exploitation and development, the region encompassing the headwaters of the Yellowstone River was set aside ". . . and withdrawn from settlement, occupancy, or sale under the laws of the United States, and dedicated and set apart as a public park or pleasuring-ground for the benefit and enjoyment of the people."

The exact origin of the name "Yellowstone" is uncertain; it either comes from the yellow, altered rocks found in the Grand Canyon of the Yellowstone River in the park, or from yellow rocks near the confluence of the Missouri and Yellowstone rivers in eastern Montana and North Dakota.

Yellowstone Park, like the Teton Range, is first and foremost a spectacular geological phenomenon. The forests, waterfalls, canyons, geysers, and wildlife of the park exist as they do because of geological processes operating deep beneath the park in the crust and mantle.

Yellowstone Park is the culmination of several geological processes that have worked through time to produce the pre-

Map showing where geysers occur in other parts of the world besides Yellowstone Park.

sent landscape. These processes may be subdivided into two categories: surficial processes that affect the surface of the land, and interior processes that are driven by heat and pressure from within the Earth. Basically, surficial processes tend to erode the land surface, while internal processes uplift the land through volcanism and mountain building. Let's review each category in light of what you will see in the park.

Surficial Geological Processes

The main surficial geological processes operating in the park are hydrothermal, glacial, and fluvial. We include hydrothermal features with surficial processes because geysers and hot springs are near-surface features that depend on a supply of ground water, although the source of heat comes from deeper within the Earth.

Hydrothermal fields throughout the world are invariably associated with recent volcanic activity, and Yellowstone is no exception. Yellowstone's hydrothermal processes, as manifested by the countless hot springs, pools and geysers, are unsurpassed anywhere in the world for their sheer number and variety. There are between 2,500 and 10,000 thermal features

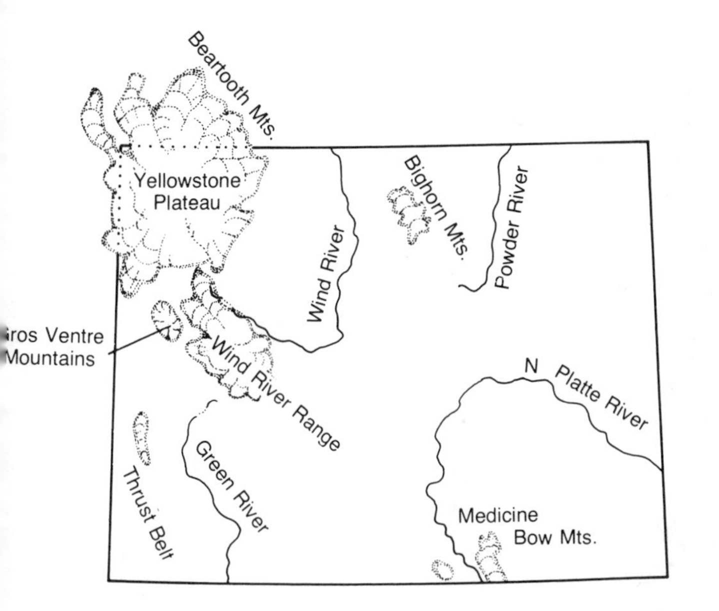

Areas of maximum extent of glaciers in Wyoming during the last Ice Age. Note extensive ice cap over Yellowstone and Wind River Mountains.

–Adapted from R.M.A.G. *Geologic Atlas* (1972)

in the park, depending on how many small ones are counted. Most thermal features are in the geyser basins, which are controlled by fractures and faults. These geyser basins are first seen as rising clouds of steam, although the amount of steam you will see is a function of the outside air temperature. Enormous clouds of steam rise on cold days, making Yellowstone especially beautiful in the winter.

Three ingredients are necessary for hydrothermal activity. First, a source of heat: The Earth beneath the park is abnormally hot due to heat conduction from cooling igneous rocks. Second, a source of water: The annual snowpack and rainfall in the park provide an ample supply. Third, a plumbing system of fractures to carry the water from the surface down to the hot rocks, then back up to the surface to form a hot spring or geyser. This works much like the radiator of a car as it picks up heat from the engine and circulates the water.

Glacial processes were also very important in producing the present landscape of the park. Yellowstone Park was affected by three major glaciations in the last 250,000 years or so, just yesterday in terms of geological time. The youngest was the Pinedale glaciation from 70,000 to 15,000 years ago; the next oldest was the Bull Lake glaciation from 200,000 to 130,000 years ago; and the oldest was pre-Bull Lake glaciation (sometimes called the "Buffalo" glaciation) which occurred around 200,000 years ago. The Buffalo and Bull Lake glaciations oc-

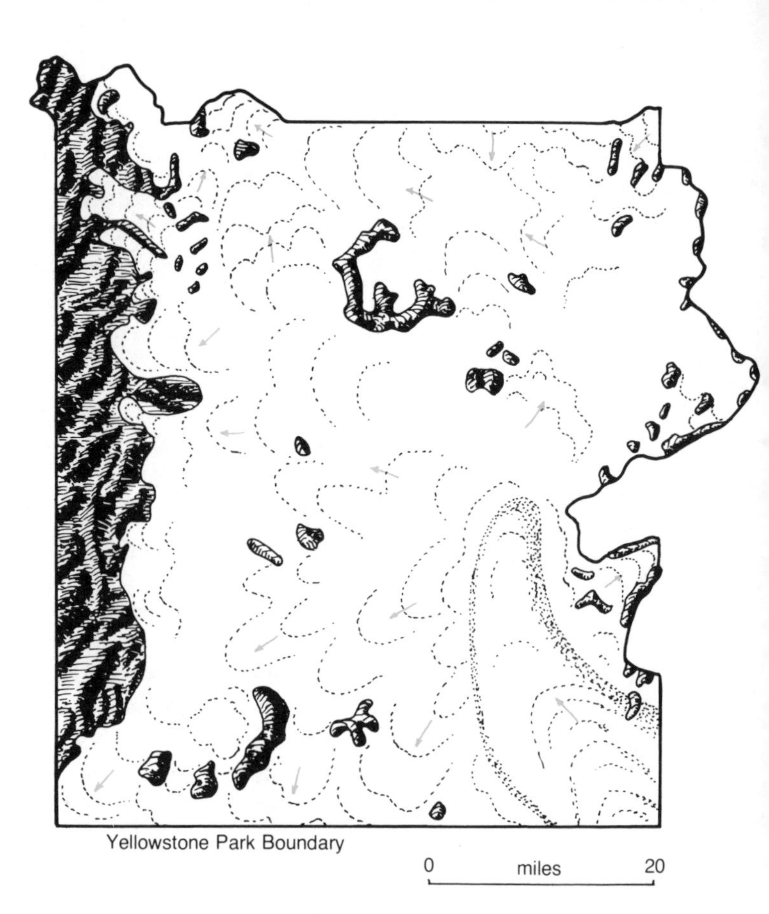

Extent of glacial ice in Yellowstone during the maximum extent of Pinedale Glaciers (about 15,000 years ago). The arrows show direction of ice movement. As shown in the diagram, some of the high peaks were not covered by ice.
–Adapted from Keefer (1976, p. 60)

curred during eruptions of the Plateau rhyolite lava flows, so most of these glacial effects on the land have been masked, but they are known from small, scattered glacial deposits or moraines. Pinedale glacial ice covered 90 percent of the park and its effects are widespread and easily recognized. The source for Pinedale ice was the high Absaroka and Beartooth ranges along the eastern margin of the park. Ice from this region flowed and coalesced into a large "ice cap" centered over Yellowstone Lake that was as much as 3,000 feet thick! Ice slowly flowed away, like giant rivers, from the park, going north down the Yellowstone and Gallatin river drainages, east down the Shoshone and Clarks Fork drainages, and south into Jackson Hole. In the Hayden Valley, lake sediments indicate that there was a dam of glacial ice across the Yellowstone River (probably near the Upper Falls); flood gravels near Gardiner, Montana, testify to the fact that this ice dam broke more than once and sent torrents of water cascading down the Yellowstone River.

This must have greatly upset any resident fish! By 15,000 years ago the Pinedale ice cap had melted away, and by 6000 years ago all glaciers in the park had melted. There are no active glaciers in Yellowstone today, although the Teton Range to the south still harbors a few small glaciers in deep, shaded canyons.

Fluvial geological processes are those that involve running water. The many spectacular waterfalls that add sparkle to the scenery are the result of rivers and streams flowing across resistant ledges of volcanic bedrock. Most famous are the Upper and Lower falls in the Grand Canyon of the Yellowstone River. A river must have enough gradient to accomplish the spectacular canyon cutting in the park. Yellowstone is a high blister, uplifted by heat, astride the backbone of the Rockies. What is the cause of this heat and resulting high elevation? To understand, we turn to the internal geological processes fundamentally responsible for the origin of Yellowstone Park.

Internal Geological Processes

Yellowstone National Park is born from the fires of volcanoes. Violent volcanic activity, more than any other geological process, is responsible for the creation of Yellowstone. There have been two main episodes of volcanism in the Yellowstone area, widely separated in time.

The oldest period of volcanism formed the Absaroka volcanic field in the Absaroka Mountains along the eastern side of the park, the Washburn Range Volcano in the north-central part of the park, Bunsen Peak south of Mammoth, and intrusive igneous rocks in the core of the southern Gallatin Range. These rocks range in age from 40 to 55 million years old; which is comparatively young in terms of geologic time.

Volcanic mudflows or lahars composed of andesite and dacite fragments spread over the future site of Yellowstone, covering ancient forests in their path. They must have resembled the mudlfows created by the 1980 eruption of Mount St. Helens in Washington. The spectacular petrified trees of Specimen Ridge in northern Yellowstone were buried under this volcanic de-

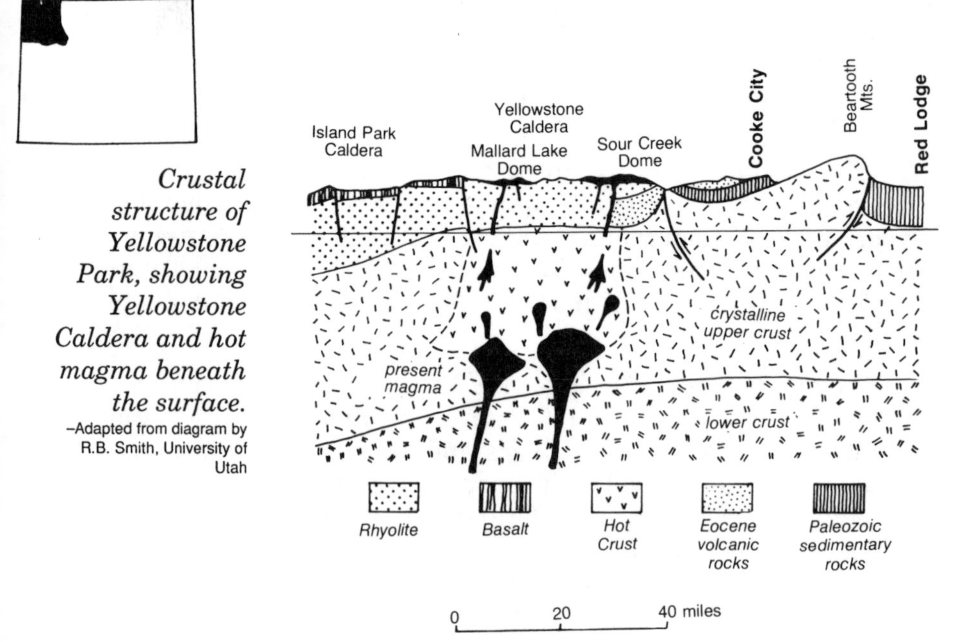

Crustal structure of Yellowstone Park, showing Yellowstone Caldera and hot magma beneath the surface.
–Adapted from diagram by R.B. Smith, University of Utah

bris, ash, and dust from nearby volcanoes. Fossilized remains of sycamore, walnut, magnolia, chestnut, oak, redwood, maple, and dogwood trees have been identified in 27 different forest layers at Specimen Ridge. The kinds of trees show that the climate was considerably more temperate then now, and that volcanism in the Absarokas was intermittent over several million years. The Absaroka episode of volcanism ended about 40 million years ago and left a gently rolling volcanic plateau, 1,200 to 2,000 feet above sea level, dotted with higher volcanic cones.

The most spectacular event in the history of Yellowstone began about two million years ago and continues to the present. This second volcanic episode, which formed the present Yellowstone Volcanic Plateau, consisted of 3 cycles of extremely violent, caldera-forming explosions. These eruptions were so large and cataclysmic that they are hard for us to comprehend. Certainly nothing remotely approaching them has occurred in recorded human history—the eruption of Mount St. Helens was trivial by comparison. Each cycle began and ended with long periods of episodic eruptions of rhyolitic lava, climaxed in between with an explosive, caldera-forming eruption of rhyolite ash flow tuffs. These violent eruptions probably lasted only a few hours or days, but each one dramatically altered the landscape of northwestern Wyoming.

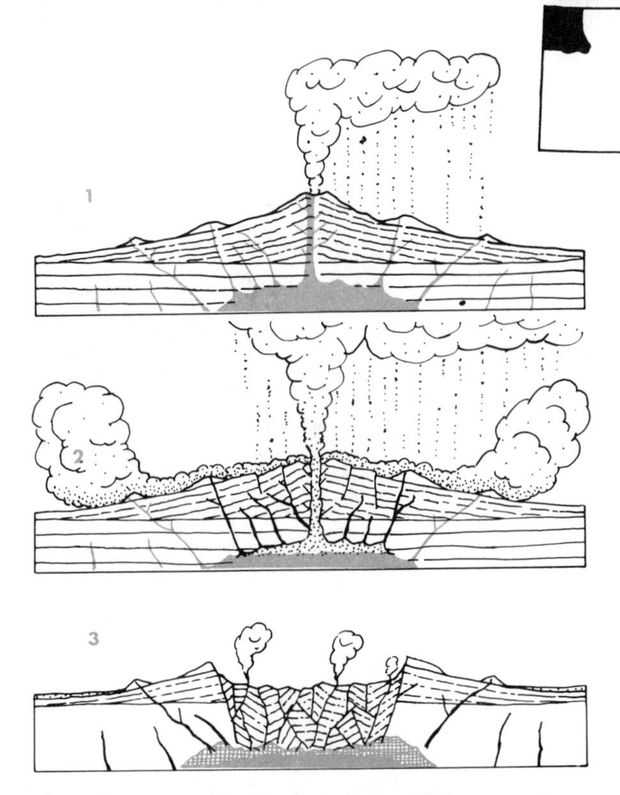

Diagrammatic evolution of a typical caldera, modeled after Crater Lake, Oregon. First, molten magma stood high in the volcano system; later it sank as eruption removed material. Finally, the volcano collapsed into its chamber, leaving a caldera, enclosing Yellowstone Lake. –Adapted from Williams (1942)

The first eruption that occurred about 2 million years ago blew the Huckleberry Ridge tuff from the giant Yellowstone Caldera. It was followed by the eruption of the Mesa Falls tuff 1.3 million years ago from the Island Park caldera. That, in turn, was followed by eruption of the Lava Creek tuff 600,000 years ago from the current Yellowstone Caldera in the central part of the park.

Each explosion produced a caldera, or giant collapse depression, that later filled with rhyolite and basalt lava flows. Yellowstone Lake occupies the depression produced by the most recent collapsed caldera, the Lava Creek tuff eruption. The rim of this caldera is marked by the steep terrain east of Yellowstone Lake, the Washburn Range north of the lake, and the cliffs adjacent to the Gibbon River. Other evidence of caldera formation include "resurgent domes" that formed at Old Faithful in the northeast corner of the caldera, possibly due to the pressure of magma pushing up from below.

Fissure eruptions of thick, viscous rhyolite lavas occurred from 600,000 to about 60,000 years ago. Collectively, there are over 30 individual flows grouped into the "Central Plateau

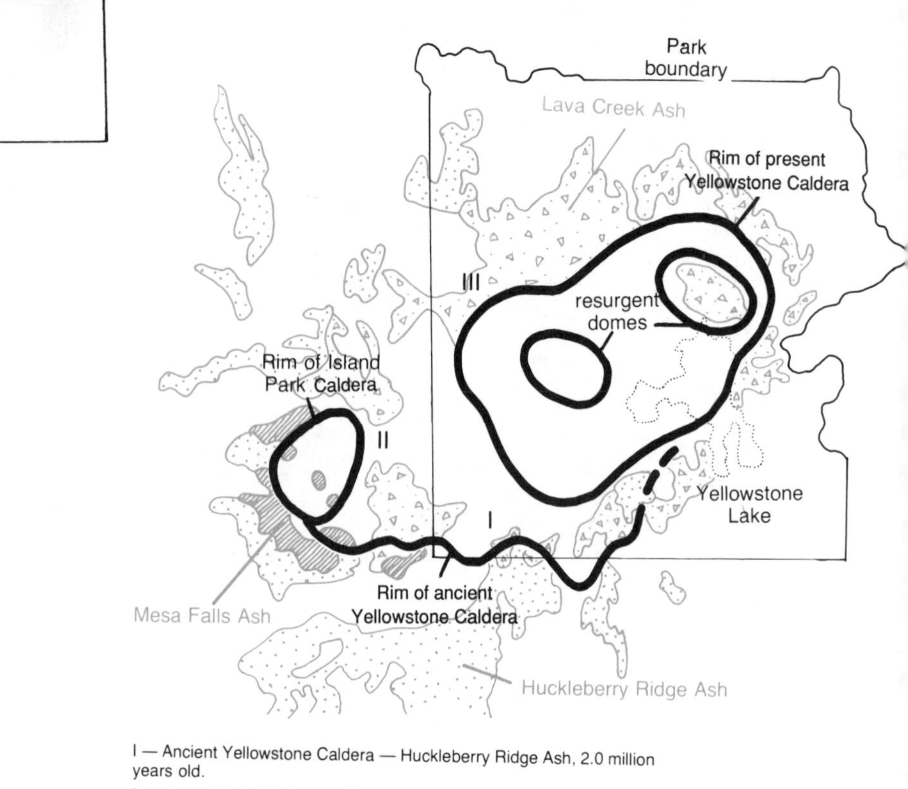

I — Ancient Yellowstone Caldera — Huckleberry Ridge Ash, 2.0 million years old.

II — Island Park Caldera — Mesa Falls Ash, 1.3 million years old

III — Present Yellowstone Caldera — Lava Creek Ash, 600,000 years old

–Adapted from Smith and Christiansen (1980)

Rhyolite Sequence" that gradually filled the floor of the most recent caldera. This event produced the gently rolling lowland terrain characteristic of central Yellowstone; in some cases, modern streams follow natural channels between the young rhyolite flows. Thus, the present topography of Yellowstone is mainly a result of the last caldera-forming volcanic eruption.

Why has all this violent volcanic activity occurred here, and will it continue? The answer seems to be related to the origin of the Snake River Plain of southern Idaho. The rocks suggest that the Snake River Plain is the trace of former Yellowstones! The Snake River Plain is a rift in the crust that has localized volcanic activity for millions of years between Twin Falls and Idaho Falls. Craters of the Moon National Monument are a recent expression of this volcanism, erupting within the last 15,000 years. Volcanism has gradually migrated eastward along the Snake River Plain, and Yellowstone Park is the most recent expression of this trend. Following earlier caldera eruptions to the west, the crust cooled and subsided and the Snake

Relation of young volcanic and tectonic features of Yellowstone to rest of northwestern US. Black areas are older Cenozoic volcanic rocks. Red areas are volcanic rocks 2 million years or younger in age.
–Adapted from Smith and Christiansen (1980)

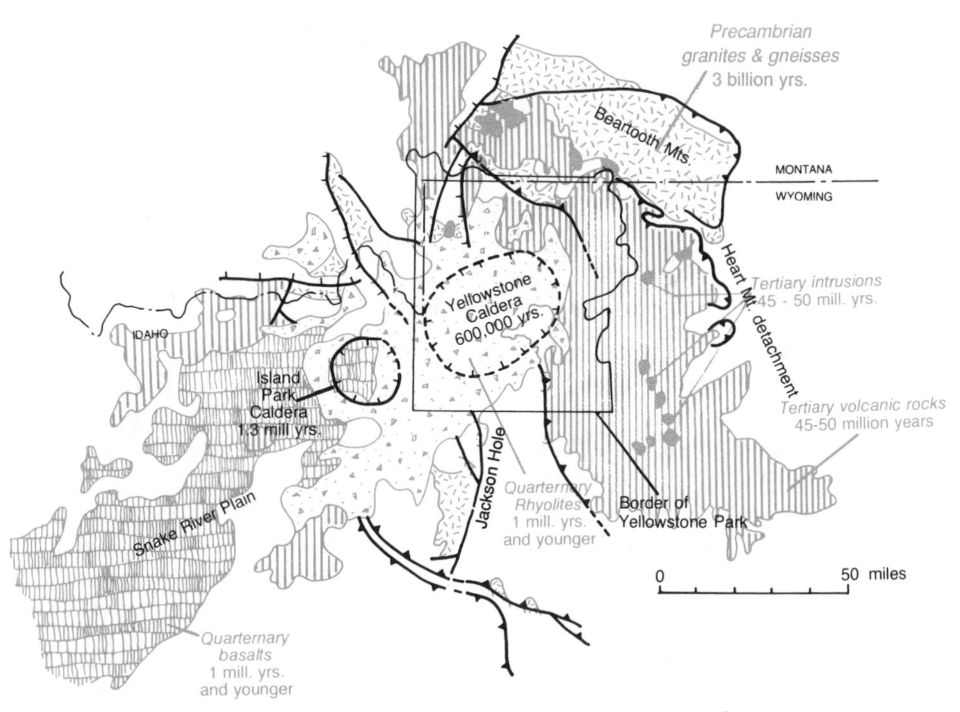

Geologic map of volcanic rocks around Yellowstone Park. –Adapted from Smith and Christiansen (1980)

River Plain was covered with flat-lying basalt lava flows. Therefore, we can predict that the present site of Yellowstone Park will eventually look like southern Idaho and a new Yellowstone will form somewhere to the northeast. Don't worry, it will be hundreds of thousands, even millions of years in the future!

North Entrance—Mammoth Hot Springs
5 mi./8 km.

Gardiner, Montana is the north entrance to Yellowstone National Park and is marked by the confluence of the Yellowstone and Gardner rivers (yes, Gardner River is spelled differently than Gardiner, Montana). The name comes from Johnson Gardner, a trapper who visited the area in 1830-31; the "i" was later added to the name of the town.

To the west of Gardiner rise two beautiful peaks, Electric Peak, the highest point in the Gallatin Range at 10,992 feet, and Sepulcher Mountain at 9,652 feet. Both contain Eocene igneous rocks formed during eruptions of the Absaroka volcanic field around 45 million years ago, at the same time the fossil forests were buried on Specimen Ridge. Electric Peak is a glacially carved "horn" that was named in 1872 by Henry Gannett, a surveyor for the U.S. Geological Survey of the Territories, who encountered a severe electrical storm while ascending the mountain. Sepulcher Mountain is named for a distinctive tomb-like rock near its summit.

The large stone arch at the north entrance to Yellowstone was built of local columnar basalt by the Army Corps of Engineers who were in charge of developing a road system for the Park. President Theodore Roosevelt dedicated it in 1903. Between the north entrance and Mammoth, the road follows the Gardner River as it cuts through drab-grey, upper Cretaceous marine shales and sandstones which form the high cliffs. These sedimentary rocks were deposited approximately 90 million years ago in a shallow marine basin which covered the west-central part of North America, long before volcanic activity began in the Yellowstone area.

The high ridgeline east of Mammoth is Mount Everts. It is composed of Cretaceous shales and sandstones, deposited in horizontal

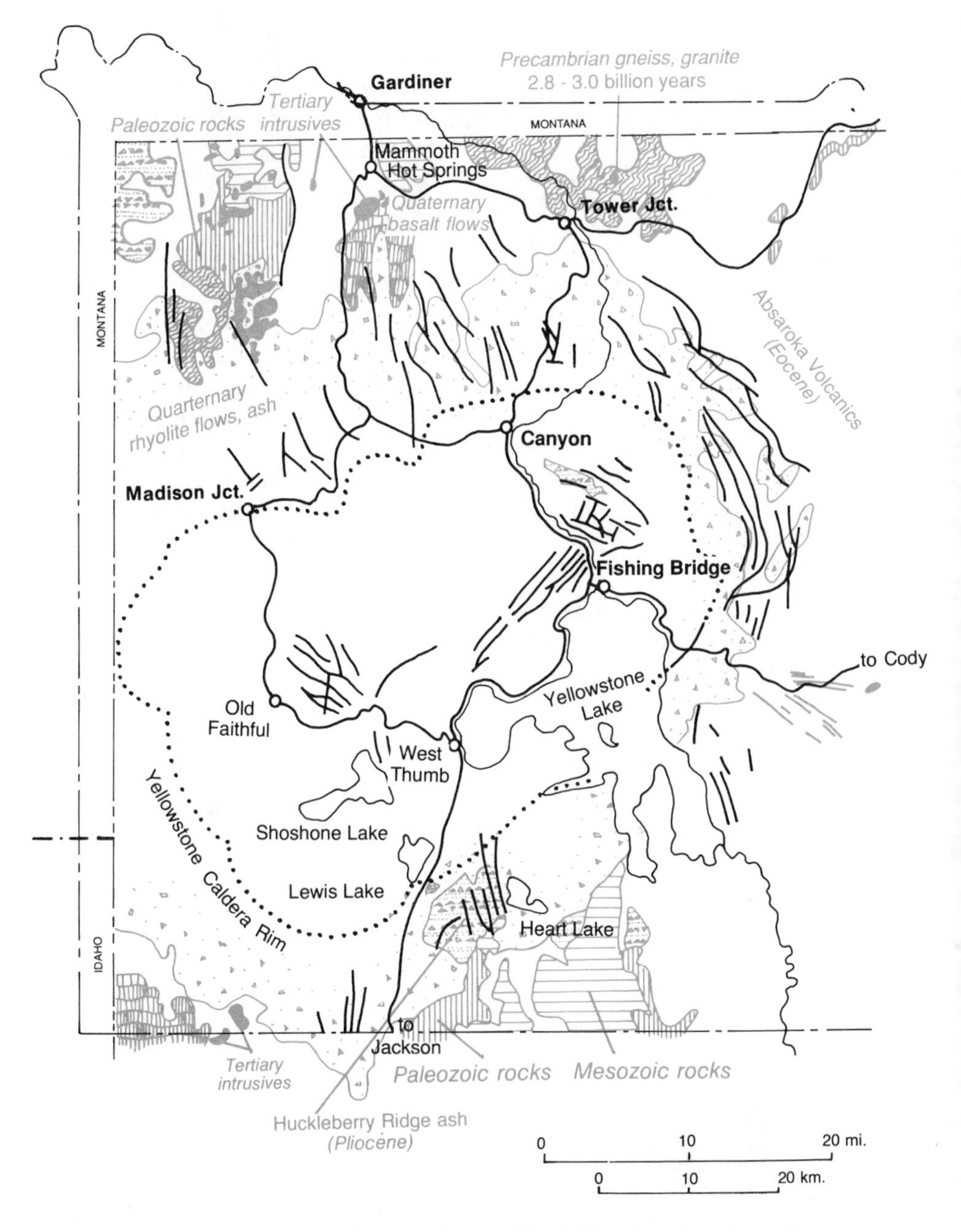

Geology of Yellowstone, showing faults (dark lines), volcanic rocks and caldera structure. –Adapted from Smith and Christiansen (1980)

Travertine terraces at Mammoth Hot Springs.

layers at the bottom of an inland seaway almost a hundred million years ago. After deposition, the strata were raised and tilted. Following a long period of erosion, the strata were then covered by the Huckleberry Ridge, tuff which erupted from the first Yellowstone Caldera around 2 million years ago. The Huckleberry Ridge tuff now forms the resistant cap rock along the summit ridge of Mount Everts. Geologist call this relationship an "angular unconformity;" approximately 90 million years spans the junction of these two rock units!

The Mammoth Visitors Center features a display of the first geological expedition to the area, led by Ferdinand V. Hayden in 1871-72. Artist Thomas Moran and photographer William H. Jackson accompanied Hayden. Their paintings and photographs, some of which hang at the Visitors Center, helped convince Congress to establish Yellowstone National Park.

Mammoth Hot Springs is an excellent example of travertine terrace development. As previously mentioned, heat, ground-water, and

Geology of the formation of Mammoth Hot Springs. –Adapted from Mammot Hot Springs Area pamphlet, Yellowstone National Park

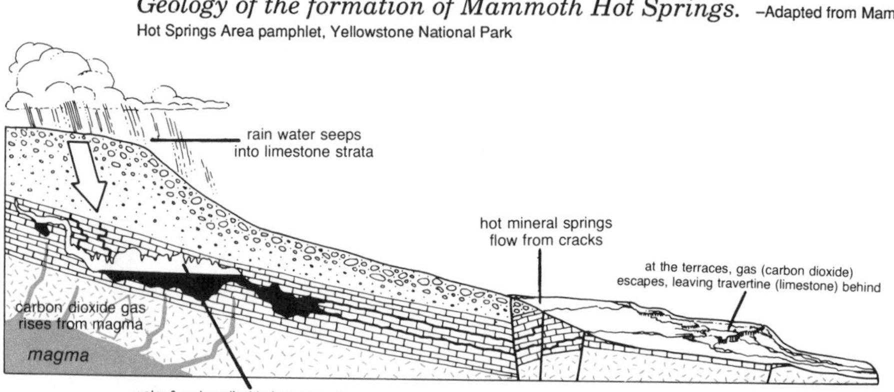

fractures are the necessary ingredients for thermal features. In addition, hot water obtains mineral ions in solution which precipitates at the surface to form terraces and mounds. The following sequence of events at Mammoth produces the terraced hot springs deposits: 1) water from rain and snow seeps into the ground; 2) the groundwater flows along fractures and porous rock where it is heated to 170 degrees F (73 C) and picks up carbon dioxide in solution, which makes the water into carbonic acid; 3) as the hot, acidic groundwater percolates through buried limestone beds, it dissolves the rock and picks up lime (calcium carbonate) in solution; 4) the lime-bearing hot water eventually reaches the surface where a sudden decrease in pressure causes the carbon dioxide to come out of solution, which makes the lime less soluable. As a result, carbon dioxide is liberated to the atmosphere and lime is precipitated as travertine. Thermal waters also may contain sulfur, producing the rotten-egg smell associated with most hot springs. Geologists estimate that the waters of Mammoth Hot Springs carry more than 2 tons of dissolved lime to the surface each day, at a flow rate of 500 gallons per minute! In effect, Mammoth Hot Springs is a giant cement factory—about 8 inches of rock are deposited each year!

Liberty Cap, at the base of the hot springs, is a single cone of travertine 37 feet high and 20 feet in diameter. It was an active hot spring until the cone became so high that hot water could no longer spill out of the top.

There are no geysers in the Mammoth area because the water here is not hot enough. A hole drilled by the U.S. Geological Survey found water temperatures in Mammoth to be about 167 degrees F. In comparison, at Norris Geyser Basin, the park's hottest area, surface water temperatures reach 199 degrees F, while water underground registered in excess of 450 degrees F!

Mammoth Hot Springs—Norris
21 mi./34 km.

Heading south from Mammoth to Norris, the road climbs steeply uphill through old travertine terraces created by former hot springs. Large, angular blocks of travertine fallen from cliffs above the road have accumulated in a huge dishevelled pile known as the "Hoodoos" or "Fallen City." These blocks cover the bedrock of Cretaceous shale.

"Golden Gate" Cliffs formed by the Huckleberry Ridge tuff (2 million years old) just south of Mammoth Hot Springs.

At Rustic Falls and the Golden Gate, the road is perched on the side of a cliff composed of Huckleberry Ridge tuff, erupted during Yellowstone's first explosive cycle around 2 million years ago. The tuff is a layered volcanic ash flow composed of pumice fragments, quartz and feldspar. Immediately east of the highway, Bunsen Peak rises to an elevation of 8,564 feet, an intrusive "plug" of dacite igneous rock that formed during the Absaroka volcanic episode of about 50 million years ago. Bunsen Peak was named for the German physicist who developed the "Bunsen Burner" and also did early research on geysers.

Shortly after passing Golden Gate, the Swan Lake pull-out offers an excellent panorama of the southern Gallatin Range and the upper Gardner River valley. The Gallatin Range is composed of Precambrian igneous and metamorphic (basement) rocks overlain by Paleozoic and Mesozoic sedimentary rocks. These rocks have in turn been intruded by igneous rocks similar in age and composition to those in Bunsen Peak. The entire Gallatin Range later rose and tilted along a large normal fault, the other side of which dropped to form the broad valley presently occupied by Swan Lake and the upper Gardner River.

Sheepeater Cliffs, named after early Indian inhabitants of Yellowstone, rise east of Swan Lake, clearly visible from the road. They are composed of Pleistocene lava flows, basalt.

Obsidian Cliffs, about midway between Mammoth and Norris, are well worth a stop. Originally called "Glass Mountain" by the famous mountain man Jim Bridger, the cliffs are indeed composed of volcanic glass, black obsidian. Obsidian displays conchoidal fracture when it

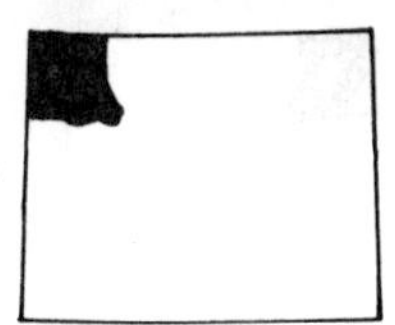

breaks, similar to broken glass or a chip on your windshield. Obsidian from this locality was traded extensively by the Indians and used to make arrowheads and other sharp-edged implements. Remember, this is a national park, so rock sampling is not permitted.

Between Obsidian Cliff and Norris, the road winds through a canyon cut into the Lava Creek tuff, a volcanic ash flow that was erupted out of the central part of Yellowstone 600,000 years ago.

Norris—Madison Junction

14 mi./23 km.

The Norris Geyser Basin is the hottest and most active in Yellowstone National Park. It constantly changes as new thermal features suddenly appear and then disappear in a few days or weeks.

In a nutshell, the Norris Geyser Basin is a very hot blister on the surface of Yellowstone. You may recall from the introduction to Yellowstone that the necessary ingredients for a geyser basin, like Norris, are hot rocks in the subsurface, an ample supply of groundwater, and a plumbing system of fractures to circulate the water through the hot rock like a car's radiator. As the water boils to the surface, minerals precipitate out of solution due to the decrease in pressure. At Mammoth, travertine (calcium carbonate) or lime is being deposited because the hot water circulates through older limestone beds underground, but at Norris a rock called sinter or geyserite, a form of quartz, precipitates because the water is circulating through volcanic rocks that contain silica.

The extremely hot water and unstable ground surface lead the park to require all visitors to stay on the walkways. Try to protect glasses or camera lenses from geyser spray. Once that mineralized water evaporates, the remaining silica is almost impossible to remove.

From Norris to Madison Junction the road drops down into the beautiful Gibbon River Canyon. Points of interest include Beryl Spring and Gibbon Falls. Rocks west of the canyon are mostly the Lava Creek tuff, while those to the east and south are mostly young rhyolite lava flows. As you drive south, you are descending into a giant collapsed volcanic caldera formed by the eruption of the Lava Creek tuff 600,000 years ago. Gibbon Canyon is on the northwest flank of this large collapsed depression that forms central Yellowstone. This is one of the largest recent volcanic features in North America! Remember, recent volcanism is the furnace that provides the heat for all the geyser and hot spring activity in Yellowstone.

Norris—Canyon
12 mi./19 km.

The road from Norris to Canyon is a short 12 mile hop that, geologically, can be divided into two segments. The western half, from Norris to Viriginia Cascade, is through ash flows of the Lava Creek tuff (600,000 years old). The eastern half crosses the Solfatara Plateau, which is composed of young rhyolite lava flows. This road, like the road from Norris to Madison Junction, crosses the northwest margin of the giant, collapsed Yellowstone caldera, formed by the eruption of the Lava Creek tuff.

Mammoth—Tower Junction
18 mi./29 km.

The road from Mammoth to Tower takes you from the Gardner River Canyon across the Blacktail Deer Plateau to the Yellowstone River. After leaving Mammoth, you cross the Gardner River on a high bridge and proceed eastward to Undine Falls in Lava Creek Canyon, a tributary to the Gardner River. A variety of rocks crop out within the canyon, but mostly you will see the 600,000 year old Lava Creek ash flow tuff, Cretaceous marine-deposited shales and sandstones (north side of the canyon), and some very young basalt lava flows called the Osprey basalt. Undine Falls formed where the creek tumbles over layers of these resistant volcanic rocks.

From Blacktail Deer Creek eastward, the road climbs over the plateau of the same name. This rolling high country is underlain by older volcanic rocks that belong to the Absaroka volcanic field, 45 to 50 million years old. They are covered by surficial sands and gravels. As you drop towards the Yellowstone River, the road winds through a variety of volcanic deposits of Absaroka vintage.

A side-road leads to the Petrified Tree one and a half miles west of Roosevelt Lodge. This tree was buried by volcanic ash and debris about 45 million years ago, exactly where it stood. Over time, it was fossilized—the organic matter of the tree was replaced by silica-rich minerals, dissolved in the ground water. The minerals exactly duplicated the cell structure of the tree, thus preserving tree rings and other delicate features. This tree was a redwood like those growing today along the California coast. Photos and reports from the late 1800s show that there were several other stumps and logs in the area, but they have all disappeared at the hands of collectors. This is the only petrified tree in the park that can be viewed from a car, although entire forests of petrified trees are preserved on Specimen Ridge.

Northeast Entrance—Cooke City—Tower Junction

29 mi./47 km.

Northeast Yellowstone is one of the most beautiful parts of the park. This area is not as famous as Old Faithful or as awesome as the Grand Canyon of the Yellowstone, but the high mountains create a rugged wilderness environment that is lost in the flatter and more crowded parts of the park. The road from Cooke City, Montana, to Gardiner is the only road in Yellowstone that is kept open year-round. The Beartooth Highway between Cooke City and Red Lodge, Montana, is one of the most breathtaking roads in North America. It is usually open from June 1 to October 15.

From the northeast entrance the road proceeds southwest along the valley occupied by Soda Butte Creek. The spectacular mountains on both sides of the valley, such as Druid Peak, The Thunderer and Mount Norris, are composed of volcanic mudflow conglomerates of the Absaroka volcanic field, about 50 million years old.

Stop to see Soda Butte, a small, isolated, cone-shaped hot springs deposit composed of travertine. The butte was formed by processes similar to those at the larger Mammoth Hot Springs. Activity is now minimal, but enough hydrogen sulfide gas is released to produce the rotten egg smell.

Soda Butte Creek flows into the larger Lamar River, which joins the Yellowstone River near Tower Junction. Bison and elk abound in this area and are often seen in large herds in the fall, winter and spring months.

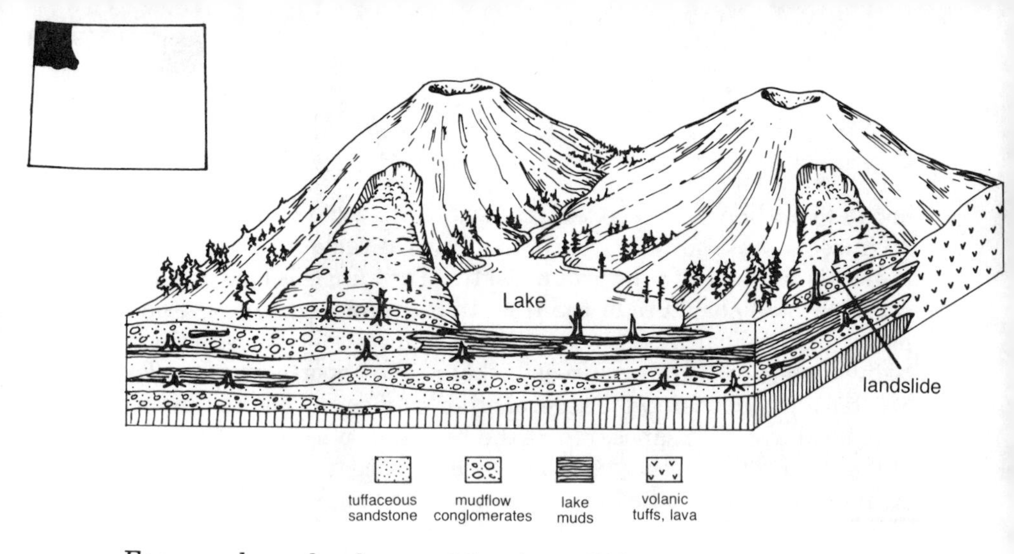

Eocene volcano landscape of Specimen Ridge. –Adapted from Yuretich (1984, p. 162)

The Lamar River valley is an excellent example of a broad, glaciated, alpine valley. The surrounding high peaks and river valleys were extensively glaciated in the recent geological past. Only 15 to 20 thousand years ago the highway would have been under hundreds of feet of ice! The large, isolated, erratic boulders littered across the valley floor were carried by glacial ice, and the swampy ponds and lakes now sit in depressions or "kettles" formed by large chunks of ice that broke off the glaciers, were covered with sediment, and slowly melted. Also, numerous landslides are visible along the hillsides, where incompetent volcanic rocks slumped into the valley after the glacial ice retreated.

The high forested ridge southwest of the Lamar River is Specimen Ridge, famous for petrified trees that were buried by volcanic

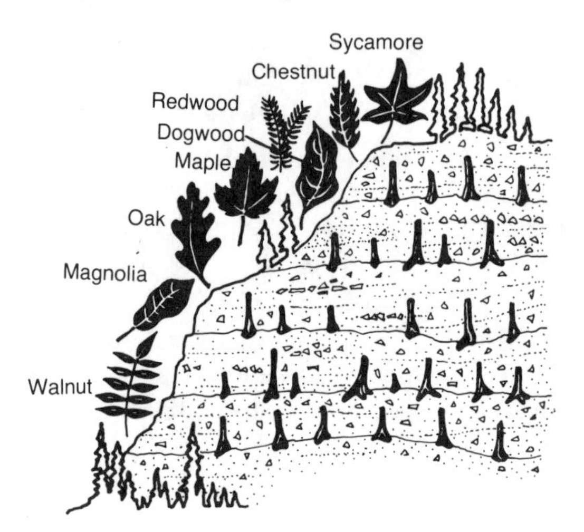

Sketch of Specimen Ridge in Hayden Valley. Burial of Eocene trees (50 million years ago) by streams and volcanic material left some standing, others prone. In the multiple layers are preserved the tree types shown at left.

Spectacular columnar jointing in Pleistocene basalt flow south of Tower Junction. Flow overlies glacial gravel and Eocene Absaroka volcanic rock.

mudflows around 50 million years ago. The process of petrification involves precipitation of silica from circulating groundwaters into open pore spaces within the wood. Detailed studies show that some of the trees were buried by volcanic mudlows; others may have been transported short distances and deposited in lakes in a vertical position. The same process has been observed in Spirit Lake at Mount St. Helens in Washington.

In the Lamar Canyon, the Lamar River cuts downward through ancient Precambrian metamorphic rocks, exposing the oldest rocks in the park. These "basement" rocks are approximately 3 billion years old and form the continental crust upon which the younger Paleozoic and Mesozoic marine sedimentary rocks were deposited. These ancient rocks crop out across north-central Yellowstone and in the Beartooth Range.

Tower Junction—Canyon—Fishing Bridge

37 mi./60 km.

The road from Tower to Canyon crosses some of the highest terrain in the park: Dunraven Pass and the Washburn Range. This section of road is geologically spectacular because the visitor can see the overprint of younger Yellowstone volcanism on the older Absaroka volcanic field on a magnificent scale. This is the best place in the park to see how Yellowstone has evolved from an old, eroded, Eocene volcanic field to a young, active volcanic field dominated by enormous col-

Gorge of the Yellowstone River as seen from Dunraven Pass between Canyon and Mammoth.

lapsed calderas, recent lava flows, and hot springs.

Just south of Tower Junction the road enters a narrow canyon formed by young basalt lava flows that overlie glacial gravels and older Absaroka volcanic rocks. The basalt lava flows display excellent columnar jointing formed as the liquid lava contracted while solidifying into rocks. Tower Creek cuts through these basalts and plunges over the underlying Absaroka volcanic mudflows to form Tower Falls.

South of Tower Falls the road climbs up the north flank of an old volcano, Mount Washburn, which was active 50 million years ago, when the fossil forests were buried. Highway cuts show layers of volcanic rock that dip to the north on the flank of this huge cone, as well as the conglomeratic rubble that was deposited on the volcano's flank. As you might expect, the volcanic debris closest to its source is the coarsest, just as you see on the flank of this old volcano.

From the summit of Dunraven Pass you can look south over central Yellowstone Park and see the massive collapsed volcanic caldera formed by the eruption of the Lava Creek tuff only 600,000 years ago. This young volcanism took place on the site of the old Absaroka volcanic field. During collapse of the enormous caldera, the Mount Washburn volcano was cut in half; the southern half of ancient Mount Washburn is now buried beneath young rhyolite and basalt lava flows in the central part of Yellowstone Park, far below you to the south. The caldera spread before you is one of the largest in the world, covering approximately 1,000 square miles. Heat from the volcanism drives the spectacular thermal features of the park.

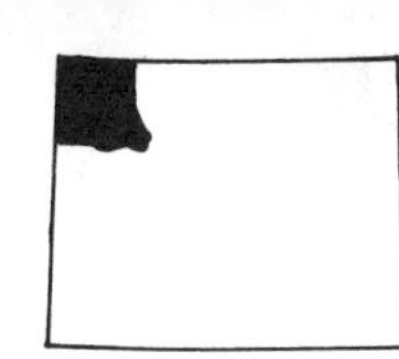

Yellowstone River Falls, Yellowstone National Park.
—Montana Travel Promotion unit

As you descend south from Dunraven Pass to Canyon, the road drops through the core of the old Mount Washburn volcano and onto the floor of the Yellowstone caldera. The sharp decrease in gradient at the south side of Dunraven Pass marks this transition. On the way down, notice the many road cuts into volcanic mudflows and breccias, the same types of rock you saw on the north side of Dunraven Pass. You have just driven through the heart of a 50 million year old volcano.

Grand Canyon of the Yellowstone River

The Yellowstone River cut this spectacular gorge during the past several thousand years. Most erosion occurs during a brief period in the spring when the snowpack melts along the Continental Divide. The upper part of the canyon, the area seen from the viewpoints, cuts through one of the Yellowstone Caldera's major fracture zones, which allowed hot water and steam to rise, altering the rhyolite lava flows. This altered rock erodes more easily than unaltered rhyolite. The spectacular colors of the canyon are the result of this chemical alteration—the normally drab brown and tan of rhyolite has been changed to brilliant yellows, reds and oranges seen in the canyon walls. This same alteration is taking place today below ground around the active hot springs and geysers throughout the park.

Resistant ledges of dark brown, unaltered rhyolite form the steep drop-offs, or "knickpoints," over which the river cascades at the Upper and Lower falls.

The road between Canyon Village and Fishing Bridge crosses the relatively flat Hayden Valley, cut into Central Plateau rhyolite lava flows. These flows flooded the floor of the caldera shortly after it collapsed. An arm of Yellowstone Lake once extended into this area and lake sediments, now covered by glacial till, blanket the valley floor. Here, the Yellowstone River meanders back and forth in its wide floodplain, creating outstanding habitat for bison, elk, and bear.

Mud pots and fumaroles abound in the Mud Volcano area because there is insufficient groundwater to create hot springs and geysers.

Between Mud Volcano and Fishing Bridge, the Yellowstone River meanders in a narrow canyon cut into rhyolite flows. Le Hardy Rapids is due to a small fault that locally increases the gradient of the river. Fishing Bridge marks the outlet of Yellowstone Lake.

East Entrance—Fishing Bridge
25 mi./50 km.

The east entrance to Yellowstone National Park is west of Cody, Wyoming, up the Shoshone River valley. The highway passes through the Absaroka Range; which extends along the eastern margin of Yellowstone National Park. The Absaroka Range is composed of Eocene volcanic rocks that range in age from about 55 to 40 million years old. The lava flows and mudflows are dominantly andesitic and dacitic in composition; that is, light-gray rocks containing about 60 percent silica. The Cascade volcanoes in the Pacific Northwest are composed of these same rock types.

The high peaks and valleys of the Absaroka Range were extensively glaciated in the recent geological past, about 200,000 years ago, as evidenced by glacially gouged valleys, sharp alpine ridges, and bowl-shaped basins called cirques on the sides of mountains. At Sylvan Pass, daily freezing and thawing of water in cracks in the cliffs during spring, summer, and fall has dislodged angular blocks of bedrock to form talus cones; a typical erosional feature in the Rocky Mountains. The cliffs at Sylvan Pass are composed of andesitic lava

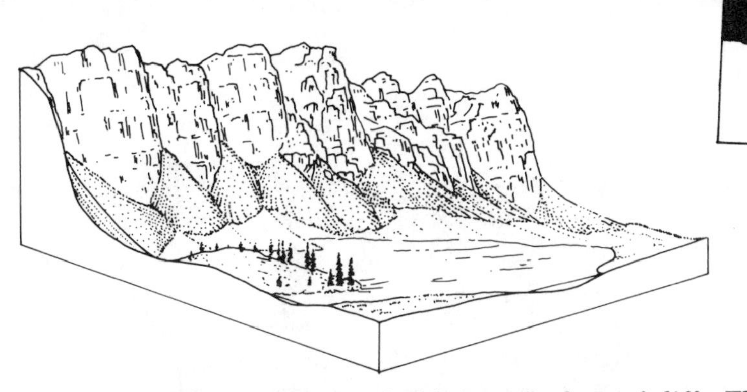

Steep talus cones are piles of rock debris at the base of cliffs. These are at Sylvan Pass. Rock fragments fall to the base, as frost action breaks them from the cliff faces. –Adapted from Hamblin (1985)

flows and sedimentary rocks derived from lava flows. West of Sylvan Pass the road descends to the level of Yellowstone Lake through more Eocene andesitic rocks. A short side trip to the Lake Butte viewpoint gives a magnificent panorama of the lake.

Yellowstone Lake occupies the eastern part of a large depression called the Yellowstone caldera, a volcanic collapse feature that formed about 600,000 years ago as an enormous explosion erupted the Lava Creek tuff. The eastern shore of Yellowstone Lake marks the eastern rim of this tremendous caldera.

Fishing Bridge—West Thumb
21 mi./34 km.

The road between Fishing Bridge and West Thumb is a scenic drive along the western shore of Yellowstone Lake. Rhyolite lava flows of the Central Plateau group lie beneath the forest west of the road, while Yellowstone Lake lies to the east. About 1,100,000-acre feet of water annually flows through the lake from the Yellowstone River—this means that the water is replaced, on the average, every 10 to 11 years.

As you drive around the lake, look closely for raised shoreline terraces that stand a few feet above the present lake level. These terraces indicate that either the lake was once higher than present or, more likely, that the land surface has been uplifted periodically perhaps due to the pressure of slowly ascending magma in the subsurface.

West thumb of Yellowstone Lake with hot spring deposits in foreground. View looking east.

West Thumb is a small semi-circular extension of Yellowstone Lake, a small caldera within the larger Yellowstone caldera. This inner caldera formed between 125,000 to 200,000 years ago during eruptions of the Central Plateau rhyolite flows. The West Thumb caldera is about 4 miles wide and 6 miles long, approximately the same size as Crater Lake Oregon, another caldera.

The West Thumb Geyser Basin lies along a narrow strip of land at the edge of Yellowstone Lake. The area features bubbling paint pots, hot springs, and a few geysers, most of them dormant. Fishing Cone, a thermal feature just offshore, was supposedly used by fishermen in the past to cook their catch directly on the hook! The practice is now forbidden and recent reports show that the water is now too cool to cook fish, anyway.

West Yellowstone—Madison Junction—West Thumb

14 mi./22 km. 33 mi./53 km.

From the park entrance at West Yellowstone, the road follows the beautiful Madison River, formed by the confluence of the Gibbon and Firehole rivers at Madison Junction. Rocks on the north canyon wall are the Mount Jackson rhyolite overlain by the Lava Creek tuff. Rocks south of the Madison River are younger rhyolite that form the Madison Plateau. Madison Junction is on the edge of the Yellowstone caldera, which dropped to the east and south during the explosive eruption of the Lava Creek tuff 600,000 years ago. The rhyolite lava flows of the Madison Plateau later erupted on the floor of the caldera. Some rhyolite flows, such as the Mount Jackson rhyolite, are older than the caldera.

At Madison Junction the road forks; one branch parallels the Gibbon River; the other follows the Firehole River to Old Faithful and West Thumb. The Firehole River cut its deep, narrow canyon through resistant layers of rhyolite called the West Yellowstone flow. Rhyolite flows are so thick and viscous, that they tend to break into rubble as they move. Many turnouts offer an opportunity to examine these rocks.

Between the Firehole River and Old Faithful, the road passes the Lower, Midway and Upper geyser basins, which include geysers, hot springs, mud pots and steam vents. Thermal activity is concentrated in this area on the flank of a highly fractured resurgent dome on the floor of the great Yellowstone caldera. The numerous fractures on and near the dome allow ground water to circulate deep into extremely hot rocks.

Old Faithful

Old Faithful has erupted "faithfully" during the many decades since its discovery, rarely missing an eruption! The following sequence of events is necessary for a geyser eruption:

1. Filling stage—the subsurface plumbing system fills with cool groundwater that is rapidly heated by the hot bedrock.

2. Splash stage—expanding steam at the bottom of the water reservoir forces the overlying water up to the surface, causing preliminary eruptions and splashing. Many visitors are tricked into thinking this is the main eruption. As water splashes out of the vent, the pressure on water at the bottom of the chamber decreases, which causes more water to flash into steam.

3. Main eruption—as large amounts of water flash into steam at depth, the overlying water is violently pushed out of the chamber in a spectacular geyser eruption.

4. Refilling—the empty water chamber again fills with cool groundwater, starting the process over again. Some geysers repeat

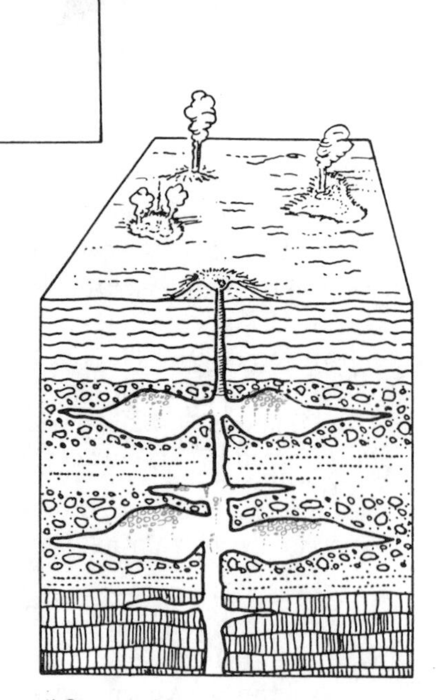

1) Steam bubbles rise in hot ground water, collect in restricted parts of geyser tubing.

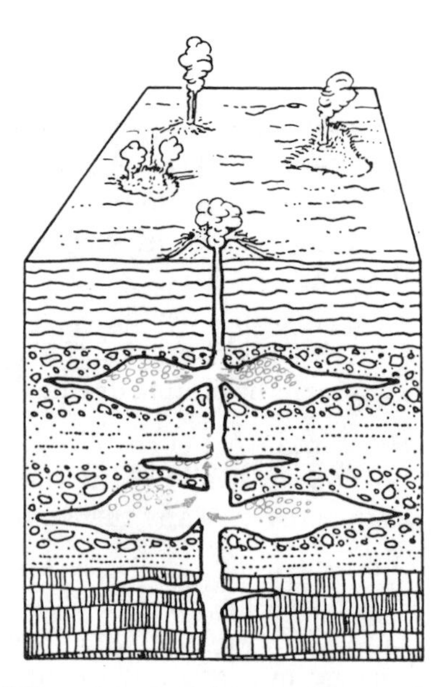

2)Expanding steam forces water upward to surface vent.

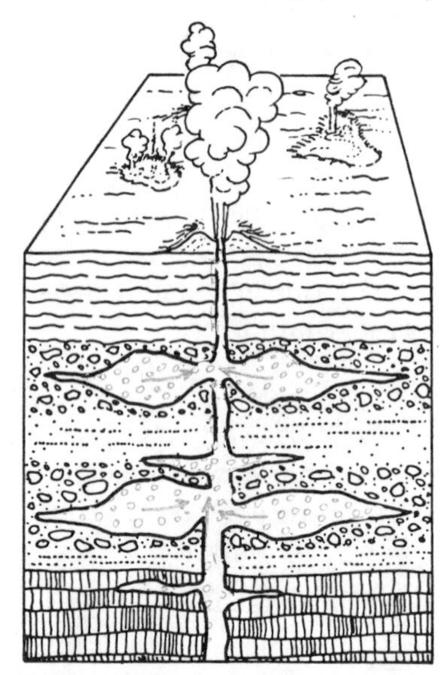

3) The preliminary water discharge reduces pressure on water at depth. The superheated water then flashes to steam and rapidly erupts upward.

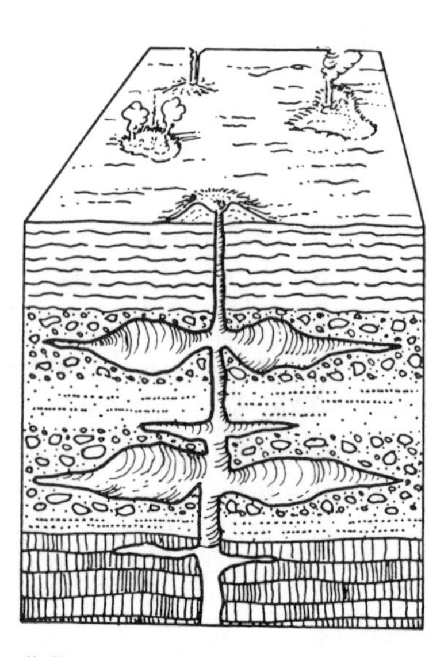

4) Eruption ceases when pressure from steam is spent and tubes empty. Water flows back in to begin cycle anew.

Eruption of a geyser. –Adapted from Keefer (1976, p. 80-81) and Hamblin (1985, p. 236)

this process with clock-setting regularity; others are sporadic and irregular. Think of a geyser eruption as a giant tea kettle that periodically fills up, boils over, then refills automatically.

Between Old Faithful and West Thumb, the road crosses the Continental Divide twice, once at Isa Lake and Craig Pass (elevation 8,262); again a few miles farther east (elevation 8,391). The Continental Divide marks the boundary between the Yellowstone-Missouri-Mississippi drainage basin and the Snake-Columbia drainage basin. It is no wonder that the Indians referred to this part of the Rocky Mountains as the "Top of the World."

South Entrance—West Thumb
22 mi./36 km.

From West Thumb to Lewis Lake the road crosses a plateau of rhyolite lava flows, the Aster Creek flow, covered with a heavy forest of Lodgepole Pine. The flows mark the southern extent of the Central Plateau rhyolite sequence that filled the Yellowstone caldera. The south margin of the Yellowstone caldera crosses the road near Lewis Lake campground. South of Lewis Lake the road descends into the Lewis River Canyon, starting with the spectacular Lewis Falls where hard ledges of rhyolite resist the erosive action of the river. Rhyolite flows of the Pitchstone Plateau crop out along the west side of the canyon and the Lava Creek tuff crops out on the east side; these volcanic rocks overlie older Paleozoic and Mesozoic sedimentary rocks that form the rugged terrain of south-central Yellowstone Park. The Pitchstone Plateau to the west is capped by one large rhyolite flow that erupted only 70 to 80 thousand years ago.

The south entrance to Yellowstone is marked by the confluence of the Lewis and Snake rivers. From here, the Snake flows into Jackson Hole, then west across southern Idaho, eventually flowing into the Columbia River in southeast Washington.

Museums In Wyoming with Geology Displays

GEOLOGICAL MUSEUM
University of Wyoming
Laramie, Wyoming

Open:
8 am-5 pm — Monday-Friday
2 pm-5 pm — Weekends

An excellent geology museum; displays on vertebrate and invertebrate fossils, rocks, minerals, geologic time, and Wyoming geology.

The Wyoming Geological Survey office is next door, where geological publications may be viewed and purchased.

GREYBULL MUSEUM
325 Greybull Avenue
Greybull, Wyoming

Open Daily:
10 am-9 pm — June-August
2 pm-5 pm — Weekends

Good fossil collection, especially large ammonites, from Cretaceous rocks in the Bighorn Basin.

SOUTH PASS CITY STATE HISTORIC SITE
South Pass City, S.W. of Lander, WY on Wyoming Highway 28

Open Daily:
9 am-6 pm
May 15-October 15

Historic mining town, 1870-1915. Buildings and displays.

GUERNSEY STATE HISTORIC SITE MUSEUM
County Road 1606,
North of US 26 about
2 mi. w. of Guernsey, WY

Open:
9 am-5 pm — Monday-Friday

Natural history and mining displays of the Sunset-Hartville mining area. Rocks and minerals. Publications.

HOT SPRINGS COUNTY PIONEER MUSEUM
700 Broadway
Thermopolis, Wyoming

10 am-8 pm — Monday-Saturday
1 pm-5 pm — Sunday
(June 1-Labor Day)
1 pm-5 pm — Daily
(Labor Day-May 1)

Geology, rocks, fossils, and minerals on lower floor. Geology of Wind River Canyon, Owl Creek Mountains, oil and gas and mining exhibits.

ANNA MILLER MUSEUM
Newcastle, Wyoming

Fossils, rocks, minerals, mining exhibits.

YELLOWSTONE PARK MUSEUM
Canyon Visitor Center
Canyon Village

8 am-8 pm — Mid June-Labor Day
8 am-5 pm — Labor Day-Mid June

Exhibits on Natural Science, geology of Yellowstone.

FISHING BRIDGE MUSEUM
Fishing Bridge
Yellowstone Lake biology and geology.

Open:
8 am-7 pm — Mid June-Labor Day
8 am-5 pm — Labor Day-Mid June

NORRIS MUSEUM
Norris Geyser Basin
Exhibits on geology of geothermal basin and geysers.

Open:
8 am-5 pm — June 1-September 30

Information Sources

Geological Society of America
P.O. Box 1719
Boulder, CO 80302

United States Geological Survey
Federal Center, Bldg. 25
Denver, CO 80225

The Geological Survey of Wyoming
P.O. Box 3008,
University Station
Laramie, WY 82071

Wyoming Geological Association
P.O. Box 545
Casper, WY 82602

The University of Wyoming
Dept. of Geology and Geophysics
Laramie, WY 82071

Suggested Reading

Creation of the Teton Landscape, by John D. Love and J. C. Reed Jr.
Grand Teton Natural History Association, Moose, Wyoming 83012

The Geologic Story of Yellowstone National Park, by W.R. Keefer
U.S. Geological Survey Bulletin 1347.

Rising from the Plains, by John McPhee
Farrar, Straus and Giroux, New York, 1986.
(Story of Wyoming and John D. Love, one of its most well-known geologists.)

Traveler's Guide to the Geology of Wyoming, 2nd edition,
by D.L. Blackstone Jr., Geological Survey of Wyoming Bulletin 67, 1988.

References

Blackstone, D.L., Jr., 1980, Foreland deformation – compression as a cause: *Contributions to Geology*, vol. 18, #2, p. 83-100.

________, 1988, *Traveler's guide to the geology of Wyoming,* 2nd edition: Geological Survey of Wyoming Bulletin 67, 130p.

Breckenridge, R.M., and Hinckley, B.S., 1978, *Thermal springs of Wyoming*: Geological Survey of Wyoming, Bulletin 60, p. 27.

Burford, A., et al., 1979, *A geologic map of Casper Mountain, Wyoming*: Wyoming Field Science Foundation, Casper, Wyoming.

Curry, Donald L., 1976, Evaluation of uranium resources in the Powder River Basin, Wyoming: *Wyoming Geological Association Guidebook.*

Dixon, J.S., 1982, Regional structural synthesis, Wyoming salient of western Overthrust Belt: *American Association of Petroleum Geologists Bulletin*, vol. 66, #10, p. 1560-1580.

Dorr, J.A., Spearing, D.R., and Steidtmann, J.R., 1977, Geological Society of America Special Paper 177, p. 65.

Eide, Ingvard Henry, 1972, *Oregon Trail*: Rand McNally & Co.

Grande, Lance, 1980, *Paleontology of the Green River formation, with a review of the fish fauna*: Geological Survey of Wyoming Bulletin 63.

Gries, R., 1983, Oil and gas prospecting beneath Precambrian of foreland thrust plates in Rocky Mountains: *American Association of Petroleum Geologists Bulletin*, vol. 67, #1, p. 1-28.

Grose, L.T., 1972, in *Geologic Atlas of the Rocky Mountain Region*, Rocky Mountain Association of Geologists, Denver, Colorado.

Hall, Charles "Pat," ed., 1976, *Documents of Wyoming Heritage*: Wyoming Bicentennial Commission.

Hamblin, W.K., 1985, *The earth's dynamic systems*, 4th ed.: Burgess Publishing, Minneapolis.

Hansen, W.R., 1965, *Geology of the Flaming Gorge area, Utah-Colorado-Wyoming*: U.S. Geological Survey Professional Paper 490, 196 p.

________, 1975, *The geologic story of the Uinta Mountains*: U.S. Geological Survey Bulletin 1291, 144 p.

Hausel, W. Dan, 1984, *Tour guide to the geology and mining history of the South Pass Gold Mining District Fremont County, Wyoming*: Geological Survey of Wyoming Public Information Circular #23.

Hausel, W. Dan, and Jones, Richard W., 1984, *Self-guided tour of the geology of a portion of southeastern Wyoming*: Geological Survey of Wyoming Public Information Circular #21.

Hoppin, R.A., 1970, Structural development of the Five Springs Creek area, Bighorn Mountains, Wyoming: *Geological Society of America Bulletin*, vol. 81, p. 2403-2416.

Houston, R.S., et al., 1978, *A regional study of rocks of Precambrian age in that part of the Medicine Bow Mountains lying in southeastern Wyoming* (with a chapter on the relationship between Precambrian and Laramide structure): Geological Survey of Wyoming Memoir #1.

Karlstrom, K.E., and Houston, R.S., 1979, *Contributions to Geology*, vol. 17, #2, front cover illustration.

Keefer, William R., 1976, *The geologic story of Yellowstone National Park*: USGS Bulletin #1347, reprinted by the Yellowstone Library and Museum Association.

Knittel, P., ed., 1974, *Field guide to the Alcova area*: Wyoming Field Science Foundation, Casper, Wyoming.

________, 1978, *Field guide to the Casper Mountain area*: Wyoming Field Science Foundation, Casper, Wyoming.

Lamar, Howard R., ed., 1977, *Readers encyclopedia of the American West*: Harper and Row Inc.

Larson, T.A., 1965, *History of Wyoming*: The University of Nebraska Press.
Love, J.D., 1970, *Cenozoic geology of the Granite Mountains area, central Wyoming (Geology of the Wind River Basin, central Wyoming)*: USGS Professional Paper 495-C.
Love, J.D., Leopold, Estella B., and Love, D.W., 1978, *Eocene rocks, fossils, and geologic history, Teton range, northeastern Wyoming*: USGS Professional Paper 932-B.
Love, J.D., and Reed, John C., Jr., 1971, *Creation of the Teton Landscape:* Grand Teton Natural History Association.
McCallum, M.E. and Mabarak, C.D., 1976, *Diamond in state-line kimberlite diatremes, Albany County, Wyoming, & Larimer County, Colorado*: Geological Survey of Wyoming, Report of Investigation #12.
McGrew, P. O., and Casilliano, M., *The Geologic History of Fossil Butte National Monument and Fossil Basin*: National Park Service Occasional Paper #3.
Mears, Brainerd, Jr., et al., 1986, *A geologic tour of Wyoming from Laramie to Lander, Jackson and Rock Springs*: Geological Survey of Wyoming Public Information Circular #27.
Miller, Daniel N., Jr., et al., 1978, *The Wyoming mineral industry*: Geological Survey of Wyoming Public Information Circular #8.
Morgan, Dale L., 1953, *Jedediah Smith and the opening of the West*: University of Nebraska Press.
Oriel, S., 1962, Main body of Wasatch formation near LaBarge, Wyoming: *American Association of Petroleum Geologists Bulletin*, vol. 46, #12.
Pierce, W.G., 1987, in *Centennial Field Guide Vol. 2*: Rocky Mountain Section Geological Society of America, p. 149.
Rocky Mountain Association of Geologists, 1972, *Geologic Atlas of the Rocky Mountain Region* (R.M.A.G. Atlas).
Royce, F., Jr., Warner, M.A., and Reese, D.I., 1975, in *Symposium on deep drilling frontiers in the central Rocky Mountains:* Rocky Mountain Association of Geologists, p. 46.
Sales, J.K., 1983, in *Rocky Mountain foreland basins and uplifts*: Rocky Mountain Association of Geologists, p. 83.
Saylor, David J., 1970, *Jackson Hole, Wyoming: In the shadow of the Tetons*: University of Oklahoma Press.
Schullery, Paul, *Road Guide for the Four-Season Road from Gardiner to Cooke City through Yellowstone National Park*: Yellowstone Library and Museum Association.
Smith, Alson J., 1965, *Men against the mountains*: The John Day Company.
Smith, R.B., and Christiansen, R.L., 1980, Yellowstone Park as a window on the earth's interior: *Scientific American*, vol. 242, #2, p. 104-117.
Spearing, Sue A., 1985, Casper: From Indians to energy, in *Wyoming Geological Association guidebook: Permian and Pennsylvanian geology of Wyoming*.
Specht, R., and Bryant, P., 1979, *Field trip road log for eastern Powder River Basin coal fields*: Wyoming Geological Association Earth Science Bulletin, vol. 12, #2.
Urbanek, Mae, 1990, *Wyoming place names*: Mountain Press Publishing, Missoula, Montana.
Wadsen, David J., 1973, *From beaver to oil*: Pioneer Printing, Cheyenne, Wyoming.
Williams, H., 1942, *The geology of Crater Lake National Park, Oregon, with a reconnaissance of the Cascade Range southward to Mount Shasta*: Carnegie Institute of Washington Publication 540, 162 p.
Wise, D.U., 1963, Keystone faulting and gravity sliding driven by basement uplift of Owl Creek Mountains: *American Association of Petroleum Geologists Bulletin*, vol. 47, #4, p. 586-598.
Yuretich, R., 1984, Yellowstone fossil forests – new evidence for burial in place: *Geology*, vol. 12, p. 162.

Glossary

Alluvial fan: a cone-shaped mass of gravel and sand deposited by a mountain stream where it runs out onto a level or nearly level plain.

Alluvium: sediments formed by rivers and streams.

Ammonites: an extinct group of shell-forming mollusks related to the living Chambered Nautilus. Both coiled and straight-shelled forms are known.

Amphibolite: a coarse grained metamorphic rock containing mainly amphibole and plagioclase minerals, with little or no quartz; forms at intermediate grades of metamorphism.

Andesite: a medium-colored volcanic rock containing a high proportion of feldspar.

Angular Unconformity: an unconformity is a substantial break or gap in the stratigraphic record that marks the absence of part of the rock record; an angular unconformity is marked by discordance between older and younger rock layers (the older rocks generally dip more steeply than the younger rocks).

Anorthosite: an igneous rock composed almost wholly of the mineral plagioclase.

Anticline: a fold that is convex upward. When eroded, an anticline has the oldest rocks in the center.

Aquifer: a porous rock layer from which water may be obtained.

Artesian well: a well in which water level rises above the top of the water-bearing layer.

Asymmetrical Fold: a fold (anticline or syncline) in which one limb dips more steeply than the other limb.

Basalt: a dark-colored volcanic rock that often contains small round vesicles or gas bubbles.

Basement: igneous and metamorphic rocks, usually of Precambrian age, lying below the sedimentary rock sequence.

Batholith: a very large mass of igneous rock (larger than 40 square miles), intruded as molten magma, often formed at least in part by melting and recrystallization of older rocks.

Bedrock: solid rock exposed at or near the surface.

Bentonite: clay formed from decomposition of volcanic ash.

Biotite: black (iron-rich) mica.

Breccia: volcanic rock consisting of fragments ejected from a volcano, lying in a fine matrix of volcanic ash.

Butte: an isolated hill or small mountain, often with a horizontal top and steep sides.

Calcite: a mineral, calcium carbonate ($CaCO_3$), the principal constituent of limestone.

Caldera: a large basin-shaped volcanic depression formed by explosion or collapse of a volcano.

Carbonaceous: containing carbon or coal derived from organic material.

Carbonate: rocks containing carbon and oxygen in combination with sodium, calcium, or other elements, paticularly as in limestone or dolomite.

Cinder cone: a small conical volcano formed by accumulation of volcanic ash and fine cinder around a volcanic vent.

Cirque: a deep, steep-walled, usually semicircular scoop in a mountain excavated by the head of a glacier.

Columnar jointing: vertically arranged, polygonal joints due to shrinkage accompanying cooling of lava and ash flows.

Competent rock: a general term for a rock that is relatively harder and or more brittle than other rocks; for example, sandstone is generally more competent than shale.

Compression: a type of deformation where the crust is laterally squeezed from 2 or more directions.

Concretion: a pebble-shaped or nodular concentration of minerals deposited around a central nucleus, usually harder than the surrounding rocks.

Conglomerate: rock composed of rounded waterworn fragments of older rock, usually in combination with sand.

Cross-bedded: with laminae slanting obliquely between the main horizontal layers of a sedimentary rock (generally sandstone).

Dacite: a volcanic rock with a high proportion of quartz and feldspar.

Detritus: a general term for particles of sediment (usually sand or silt-sized grains of quartz) produced by the disintegration and weathering of rocks on the earth's surface.

Dike: a thin body of igneous rock resulting when liquid magma intrudes a vertical joint in the rocks.

Diorite: an intrusive igneous rock composed essentially of sodium plagioclase and hornblende, biotite or pyroxene; small amounts of quartz may also be present.

Dip: the angle at which a rock layer is inclined below the horizontal.

Dolomite: a limestone-like rock containing magnesium carbonate as well as calcium carbonate.

Earthflow: a slow flow of soil lubricated with water.

Erosion: processes by which rocks are loosened or dissolved and removed from any part of the earth's surface; canyon-cutting by a stream is a common example.

Escarpment: a cliff or steep slope edging a region of higher land.

Evaporite: a mineral deposited from highly mineralized or salty water as a result of evaporation.

Exfoliation: a surface weathering process involving the spalling off of layers or concentric shells of rock from the underlying bedrock, similar to the concentric layers of an onion.

Extrusive rocks: igneous rocks that cool on or very near the earth's surface; volcanic rocks.

Facies: a body of rock as distinguished from other parts of the rock body by appearance or composition; for example, a sandstone layer may laterally grade into a shale layer, each constituting a "facies" of the layer.

Fault: a break in the rock along which rocks on either side have moved relative to each other.

Fault scarp: a cliff formed by a fault, usually modified by erosion.

Feldspar: a group of abundant light-colored rock-forming minerals.

Flatirons: triangular-shaped remains of hogback ridges steeply tilted against the flank of a mountain.

Fluvial: the processes of stream/river erosion or deposition.

Fold: geologic structures in which rock layers are buckled into wave-like forms; typically produced by horizontal compression.

Formation: a named, recognizable, and mappable unit of rocks.

Gabbro: the intrusive equivalent of basalt; composed mostly of calcic plagioclase and pyroxene minerals.

Geomorphology: a branch of geology which deals with the earth's surface features or landforms.

Gneiss: a coarse-grained metamorphic rock with alternating bands of granular crystalline minerals such as quartz and feldspar, and fine, often platy dark minerals such as biotite.

Graben: a linear, down-dropped block of the earth's crust that is bound on its margins by normal faults.

Granite: coarse-grained intrusive igneous rock with feldspar and quartz as principal minerals.

Graphitic schist: a metamorphic rock composed of layers of mica and graphite.

Ground moraine: material deposited on the ground surface by a melting glacier.

Groundwater: subsurface water filling rock pore spaces, cracks, or solution channels.

Group: a stratigraphic unit consisting of several formations, usually originally a single formation subdivided by subsequent research.

Gypsum: a common evaporite mineral ($CaSO_4$), the main ingredient of plaster.

Hanging valley: a valley whose floor is substantially higher than the floor of the valley into which it leads.

Hanging Wall: the block of rock that overlies a fault; the hanging wall moves down for a normal fault and up for a reverse or thrust fault.

Hematite: an ore of iron (Fe_2O_3).

Hogback: a sharp ridge produced by erosion of highly tilted rock layers, one of which is more resistant than the others.

Horn: a mountain pinnacle that has been produced by glacial erosion on all sides of the mountain (examples — Matterhorn in Switzerland and Grand Teton in Wyoming).

Hornblende gneiss: dark gneiss containing hornblende as the most abundant mineral.

Hydrostatic pressure: pressure caused by weight of water in water-bearing rock layers.

Hydrothermal: caused by hot water.

Icecap: glacial ice which spreads in all directions over a high, relatively flat surface.

Igneous rock: rock formed by solidification of molten rock material.

Incompetent: relatively weak and unable to support its own weight or the weight of overlying material.

Incompetent rock: a rock that is relatively less hard or less brittle than other rocks; for example, shale is generally incompetent compared to sandstone.

Injection gneiss: gneiss containing sheets and veins of granite injected under great pressure deep in the earth's crust.

Intrusive rock: igneous rock that has hardened from molten rock material (magma) before reaching the surface.

Iron formation: a general term for a metamorphic rock that contains layers of iron-bearing minerals.

Joint: a fracture in rock along which no appreciable movement has occurred.

Kaolin: a type of clay usually formed by decomposition of feldspar minerals.

Karst: a distinctive type of landscape where solution in limestone layers has caused abundant caves, sink holes, and solution valleys, often with red soil residue.

Kerogen: solid oil-like substance in oil shales.

Laccolith: a body of intrusive rock that squeezed between rock layers, doming those above.

Laramide orogeny: a mountain-building episode in the Rocky Mountain region that occurred from the late Cretaceous to early Eocene, that was characterized by the involvement of continental (granitic) crust in the faulting and folding that occurred.

Lateral moraine: an elongate moraine along the side of a valley, deposited at the side of a valley glacier.

Leucite: a mineral found in potassium-rich volcanic rocks.

Limestone: a rock composed mostly of calcite (calcium carbonate) derived from the shells of ancient invertebrate organisms.

Lode: a deposit of valuable minerals; a vein in solid rock, in contrast to placer deposits.

Magma: molten rock from which igneous rocks eventually solidify.

Magnetite: a magnetic ore of iron (Fe_3O_4).

Marble: a metamorphic rock created by baking and recrystallization of limestone.

Mesa: a tableland or flat-topped mountain or hill, usually capped by a resistant rock layer and edged with steep cliffs.

Metaconglomerate: a sedimentary rock composed of pebbles, cobbles or boulder that was metamorphosed by heat and/or pressure.

Metagraywacke: a poorly-sorted sedimentary rock composed of grains of quartz and feldspar with clay-rich matrix, that was metamorphosed by heat and/or pressure after deposition.

Metamorphic rock: rock formed from older rocks that have been subjected to great heat and pressure or to chemical changes.

Mica: a group of minerals characterized by their way of separating into thin, platy flakes, usually with shiny surfaces.

Migmatite: see *injection gneiss*.

Monocline: a fold or flexure in stratified rock in which all the strata dip in the same direction.

Monzonite: an intrusive rock containing mostly feldspar minerals, with very little quartz and few dark minerals.

Moraine: an accumulation of gravel, boulders, and dirt, deposited by a glacier.

Muscovite: white or light brown mica.

Pyroxene: a group of dark minerals common in igneous rock.

Quartz: a hard, glassy mineral, silicon dioxide (SiO_2), that is one of the commonest rock-forming minerals.

Quartzite: a metamorphic rock formed from sandstone cemented by silica.

Ramp Anticline: an anticline formed by layers of rock that are folded over a ramp or step in an underlying thrust fault (like a rug draped over a stair step).

Recessional moraine: a moraine or moraines deposited at the terminus of a glacier as it haltingly retreats (melts) up-canyon.

Redbeds: red, pink, and purple sedimentary rocks, usually sandstone and shale.

Rhyolite: light-colored volcanic rock containing invisibly small crystals of quartz, feldspar, and biotite.

Rock glacier: a glacier-like tongue of angular broken rock, usually lubricated by water and ice and moving slowly like a true glacier.

Sandstone: sedimentary rock composed of sand grains.

Scarp: cliff or steep slope.

Schist: metamorhpic rock whose parallel orientation of abundant mica flakes causes it to break easily along parallel planes.

Shale: platy sedimentary rock formed from mud or clay, breaking easily parallel to the bedding.

Silica: silicon dioxide (SiO_2), occurring as quartz and as a major part of many other minerals.

Siliceous sinter: hotspring deposits composed largely of silica.

Silicified: impregnated with silica.

Sill: a tabular sheet of igneous rock formed by hardening of magma intruded between horizontal or nearly horizontal rock layers.

Siltstone: a clastic sedimentary rock composed of silt-sized grains of quartz

Sinkhole: a large depression caused by collapse of the ground into an underlying limestone cavern.

Slickenside: a scratched and polished surface resulting from movement of rock against rock along a fault.

Stock: a medium-sized mass of intrusive igneous rock, smaller than 40 square miles at the surface.

Strata: layers or beds of rock. Singular in stratum.

Stratified: formed in layers, as sedimentary rock.

Strike-slip fault: a fault in which blocks of the earth's crust move laterally past one another, rather than vertically.

Symmetrical fold a fold in which the limbs dip at equal angles.

Syncline: a downward fold in layered rocks. When eroded a syncline has the youngest rocks in the center.

Tailings: waste debris from ore-processing mills.

Talus: fallen broken rock collected at the foot of a hill or cliff.

Tension: a type of force in which the earth's crust is pulled apart; typically produces normal faults, like the Teton normal fault.

Terminal moraine: bouldery glacial debris dumped at the lower end of a glacier at the time of its greatest extent.

Thrust fault (also reverse fault): a type of fault in which the hanging wall block (above the fault) moves up relative to the footwall block (below the fault); commonly produced by compression of the crust.

Travertine: a chemical precipitate from hot springs composed of calcium carbonate (calcite)

Triangular faceted spurs: a steep mountain front formed by very recent normal faulting that has been dissected by streams to form triangular faces or "spurs."

Tuff: a rock formed of compacted volcanic ash and cinder.

Type locality: the place at which a formation is best or most typically displayed, and from which it is named.

Unconformity: a surface of erosion that separates younger strata from older rocks.

Ultramafic sills: sheets of intrusive igneous rock that were injected parallel to bedding or foliation in the country rock, composed of ferromagnesium silicates with little or no quartz or feldspar; probably derived from the upper mantle.

Vein: a rock containing ore minerals, usually a tabular mass.

Volcanic ash: fine material ejected into the air from a volcano.

Water table: the upper surface of groundwater, below which soil and rock are saturated.

Welded tuff: volcanic ash which has been hardened by the original heat of the particles and the enveloping hot gases.

Index